Coal Mining in the East Neuk of Fife

Other earth science titles from Dunedin include:

The Abyss of Time (2015)
Paul Lyle
ISBN: 9781780460390

Volcanoes and the Making of Scotland (2015)
Second edition
Brian Upton
ISBN: 9781780460567

Geology and landscapes of Scotland (2013)
Second edition
Con Gillen
ISBN 9781780460093

For further details of these and other Dunedin Earth and Environmental Sciences titles see www.dunedinacademicpress.co.uk

COAL MINING IN THE EAST NEUK OF FIFE

John McManus

Professor John McManus FRSE
Formerly of the School of Geography and Geosciences,
University of St Andrews

DUNEDIN

EDINBURGH ◆ LONDON

Published by
Dunedin Academic Press Ltd
Hudson House, 8 Albany Street
Edinburgh EH1 3QB, Scotland

London Office: 352 Cromwell Tower, Barbican
London EC2Y 8NB

www.dunedinacademicpress.co.uk

ISBN 781780460727 (Hardback)

British Library Cataloguing in Publication data
A catalogue record for this book is available from the British Library

Typeset by Makar Publishing Production, Edinburgh
Printed in the Disunited Kingdom by CPI

Contents

List of illustrations

List of tables

Preface

My recognition of a need for modern information on the former coal industry in the east of Fife came when a farmer on a geological walk which I was leading spoke up. 'Your map has a mark on it showing a shaft top in the middle of that big field. That was exactly where we almost lost a new tractor a few years back. If we had known, we would have gone nowhere near it. Fortunately, no one was hurt'.

The comment drove home the need for a good level of information to be readily accessible for those using the land for farming, or for potential building projects, such as planners. That information exists in much of Fife but requires patient delving into old records, few of which are held in public libraries. The lack of knowledge about this formerly thriving industry, which began to go into decline in the late eighteenth century, does lead to unforeseen accidents with housing, vehicles, animals and people all vulnerable to loss of ground support. As the industry began to fail, the people who worked the mines often left the areas concerned, the buildings fell into disrepair and the memories of the sites concerned faded.

In major city developments above former mining centres, as in Glasgow, the records have been updated by the officers of the British Geological Survey, for the intense mining in the past has led to many building foundations being found wanting. Elsewhere, with less pressure for structural support, the availability of such information is of a much lower standard.

The methods used have evolved greatly through time, and the uses to which the coal has been put became highly diverse, but in that time the resource has been largely worked out.

The present text is an attempt to gather the widely dispersed knowledge of the former mine workings in the East Neuk of Fife. It attempts to explain the principal problems encountered by those seeking the coals below ground over more than eight centuries in an area where many residents are unaware of its significant role in the history of mining. The coal and oil shale extraction underwent periods of intense activity with much quieter intervening periods, principally as a result of external demands. The geographic limits of the search area are defined by the coast to the south and east, a north–south line from Lower Largo (at the head of Largo Bay) to Cupar, and in the north is defined geologically by the boundary with older rocks deposited before the formation of coals.

The preparation of a preliminary account of the nine coal mines between Ceres and St Andrews opened access to substantial records in the Cupar Library, where notable events in the East Neuk during the seventeenth, eighteenth and nineteenth centuries are systematically recorded, including the exploration, evaluation, exploitation and exhaustion of the reserves at many sites.

The Old and New Statistical Accounts provided by the parish ministers in the 1780s and 1840s, few of whom were either knowledgeable or interested in the work carried out below ground by their parishioners, generally contributed little scientifically useful information. Nevertheless, some of their comments are perspicacious. Of greater value are the accounts of the geological-mining consultants such as Landale, and later Gemmell. Much of their work was carried out before the substantial investigations of Geikie and others of the Geological Survey of Scotland, (later renamed as part of the British Geological Survey). In 1902 Geikie produced the important Memoir on the Geology of East Fife.

David Landale, mining engineer, reference to whose reports features highly in this collection, died on 27 December 1895 aged 90. His last known involvement with the Newbigging of Craighall workings was via a letter from the Factor on 5 March 1895. Among many awards for his work, he was recognized by the University of St Andrews with the award of an Honorary Doctorate of Science shortly before his death.

The papers of estate owners, often in the form of day book entries, hold a considerable amount of information, most of which is far more detailed than could be used directly here. Some of these extend back into the sixteenth century, showing sales of coals to the royal palace in Falkland and to the troops stationed in the East Neuk. There were many mining families, some at the coal face, and others moving from area to area as estate managers or oversmen (deputies). The mining diaspora is to be found throughout Fife today. Access to their records has been granted by several of the estates, but access to consultants' reports generally proved more straightforward.

To the many people who have assisted me throughout several years of searching and writing, I wish to express my deep gratitude and admiration. So many searched through documents, letters, charters and old papers in the Special Collections Section of the University of St Andrews Library, the reference collections in other libraries throughout Fife and, of course, the local museum staff, whose considerable knowledge saved me several months of endeavour. Fortunately, some added to their own knowledge of local features and history, gaining information now being passed on in their own work. The assistance of geological research and technical staff in field visits (Rosalind and Richard Batchelor) and photography of fossil samples (Stuart Alison) is very much appreciated.

The back-up team has included my son Neil, forever answering apparently simplistic questions, and daughter Kay Smith of B.G.S. with substantial assistance with IT concepts and techniques of manipulation, including GPS reference calculations. In reality, the older generation has learned greatly from the younger. Very special mention must go to Dr Paula Martin and her husband Dr Colin Martin for their extensive advice on the Pittenweem Coalfield, helping with the preparation of many text figures for publication, and permission to use my own interpretation of several of Colin's air photographs. Many of the maps and diagrams have been prepared by Graeme Sandeman of the University of St Andrews. I most especially wish to acknowledge generous financial assistance from the Mining Institute of Scotland, and similarly from the School of Geography and Geosciences towards the costs of production. Encouragement has been at an embarrassing level from the staff of the Fife Folk Museum in Ceres and the Crail Museum. The staff at National Museums Scotland and the British Geological Survey in Edinburgh have provided many insights to the past, which I doubt that I would have found unaided.

As few of the concepts and techniques of working are widely known outside the industry, I have attempted to provide visual support to the text. In this respect, I am delighted to be allowed to include a selection of the works of Derek Slater, a former miner, now a very successful art teacher, to illustrate the text. His atmospheric paintings reflect many of the activities of the miners. I am very fortunate to be able to use his work.

Throughout the last five years or so, I have received enthusiastic and unwavering support from my partner Wynne Harley, without whom the final text would not have seen the light of day. My admiration of her forbearance and patience, and my gratitude is profound. On the journey towards completion of the work, we have both learned so very much of the geology and IT together.

During the production of this work, I have freely discussed a very wide range of topics with many contacts ranging through the Goldsmiths of Largoward to Ian Terris, Dr Peter Yeoman, Dr Marie Robinson, Professor B.P. Lenman, Professor R. Crawford, Professor D.J.A. Williams, and Professor T.C. Smout, Historiographer Royal. To all of these and innumerable others not individually named, I give my most grateful thanks and earnestly hope that I will not have grossly traduced any of their beliefs. However, I recognize that inevitably there will be errors in my judgement (such as maps lacking roads and the use of GPS references) and problems remaining in the text. The errors of judgement are all mine. Finally, I wish to record my thanks to Anthony Kinahan and his supporting staff at Dunedin Academic Press for their assistance and understanding.

John McManus
November 2016

Map of Localities referred to in the text (roads not shown). (Prepared by Graeme Sandeman.)

Chapter 1

Introduction

The Kingdom of Fife has long been known as a source of coal, used locally and exported across Britain and the Low Countries for more than 400 years. Several very informative accounts of the coal mining industry in Fife have been published in the past century. The majority of these, which are listed at the back of this volume, relate principally to the collieries of Central and West Fife where the industry remained active and the winning of coal continued until well after Nationalization in 1947. Some of the coals in Central and West Fife were initially worked at the coast or along valley sides inland, usually at shallow depths, close to the land surface. Further inland the bulk of the best coals were at greater depths and often in geologically complicated structures, so many of these western coals were not worked until relatively recent times using more advanced modern methods, which gave secure access to the deeper deposits. The last active underground operations in Fife were terminated within the present century. Today there remain dwindling numbers of veteran miners who worked in the deep mines through the final days of the industry, which experienced progressive closure towards the end of the twentieth century. In such places family memories of the industry are still very vivid. By 2010 virtually the only remaining winning of coals had become confined to the open-cast operations in Central and West Fife. They too appear to be coming to an end, although there are substantial reserves of this important economic resource still in place. The present economic policies find it cheaper to import most of the raw material from Poland or Australia, despite the transport costs involved.

Long forgotten by most residents of the district and their families are the early coalfields of the east of Fife, and it is to this area that attention is directed in the present contribution. Many of the most readily accessible coastal or shallow inland deposits were to be found towards the east of the Kingdom, principally to the south and east of the Riggins, the west-trending ridge that forms the spine of high ground extending eastwards from the Lomond Hills to the sea at St Andrews. Details of several of the coalfields extending to relatively shallow depths from the surface and situated on the northern flanks of the ridge have been examined in an earlier review (McManus, 2010). In all areas of Fife, actively updated outline histories of the workings and the people involved are being recorded on the Web by Michael Martin and his webmaster associates providing a thoroughly researched, first class source of information that is highly commended to the reader. Most of the more widely published writings on the industry refer principally to the life and working conditions of the colliers and their families. The historic roles played by the mining organizations and unions in striving to improve the lot of the collier communities are fully examined by Arnott (1955). In the present account, attention is directed towards the underlying geological controls, which have not only served to provide the raw materials in abundance but have, in so many cases, severely limited the possibilities for extracting usable coals by fracturing, folding and even by burning the coals below ground through volcanic and associated activities.

Many of the collieries were active producers of coal from the early days of working the resources, but for few of these are detailed records of the development history and problems faced available today. Once newspapers began to be circulated, the details of notable events of some of the workings, such as ground breaking and drilling of boreholes, were recorded by the local press, and selections of these have been extracted by Sparling (2004, 2005, 2008) in a series of short booklets on mining activity in certain areas. However, it must be remembered that newspapers of any form did not appear until the start of the eighteenth century. Those able to read them were few in number and generally not deeply interested in matters relating to the mining industry. Particular attention has been given to the mines and people of the Largoward district in an unpublished, but well-illustrated, account by Terris (2010). A second significant source is an unpublished summary of reports on strategic distribution of potentially valuable coal deposits produced by Dott (1944) for the British Geological Survey (BGS). A third collection of reports, fully documented for the National Coal Board (NCB), was culled by Forbes (1955). These were derived principally from the works of David Landale, a leading nineteenth century mining engineer. As a consultant, he produced the first coalfield maps of Fife in 1835 and 1837 and for more than half a century recorded details of a wide range of mine workings. It is largely from the work of Landale that detailed plans, formerly held by the National Coal Board at Dysart, have survived. Following the important leads by Landale, the British Geological Survey in its many guises has undertaken more systematic mapping of the terrain, the coals and other valuable materials to be found in Fife, with the provision of foundation Memoirs by Geikie on the Geology of Eastern Fife (1902), the Geology of Central and Western Fife (1900) and more recently, the updated account of the Geology of East Fife by Forsyth and Chisholm (1977). Many subsequent detailed accounts of coastal exposures of the rock sequences, their fossil contents and detailed records of findings from exploratory boreholes have been given by the Survey officers in the Bulletins of the Geological Survey.

In this account, as with several of the earlier reports, it will be necessary to review some of the changing historical and socio-economic factors operating at the different periods during which the mining industry was developed. These impacted directly upon the collier families in terms of working conditions, hours served below ground, wages, demand for the products, restriction of demand when parliament, fearing shortages at home, banned the export of coals from the country, loss of trade through changes in national alliances, the incidence of deadly diseases, improvements to road transport for distribution of the coals, and the later introduction of railways. The coal industry did not exist in isolation.

Chapter 2

The geological setting of the East Neuk

The location of the East Neuk, on the extremity of the temperate Fife peninsula reaching into the North Sea from Central Scotland, is probably no more than a temporary one in a geological sense. The landmass of which it is a part has been moved around the surface of the Earth's globe for hundreds of millions of years. Although locally we as individuals cannot detect the very slow movement of the continent on which we sit today, there is no reason to consider the present position as anything but impermanent.

It is normally believed by geologists that in the distant past a mass of gases and interstellar dust came together to form a single body whose temperature increased with energy derived from collisions between the impacting solid particles and also from radioactive decay of elements within them. As the early molten mass of the proto-Earth increased from the collisions so the heavier materials such as iron, nickel, copper and gold sank towards its centre under the influence of a growing gravitational field, the lighter elements becoming more common outwards. The planet's core is believed to be an iron-rich solid more than 5100km below the surface, surrounded by a 2300km thick layer of other material, most probably iron. Above this lies the mantle, also largely of heavy minerals, but containing a great variety of elements. The mantle extends to within 350km below the surface. Between the mantle and crust lies the asthenosphere, the uppermost 100km or so of which is a relatively plastic layer of rock, the lithosphere, upon which rest both the continental and oceanic crusts.

Rather more than 4500 million years ago, the outer parts of the molten mass forming the planet had cooled sufficiently to permit the formation of crystals within the liquids at the surface, and these came together to give rise to the first rocks. These igneous rocks, formed by crystallization directly from a molten source, would have consisted of minerals such as olivines, pyroxenes, the basic feldspars and magnetite, very similar to those forming new oceanic crust today. Later the silica-rich feldspar minerals and quartz developed, giving rise to relatively low-density, light-coloured rocks of the granite clan, creating the first continental crust. Above the new ground surface, the gases still escaping from the Earth's interior eventually condensed producing acidic rains. The rainwaters not only served to further cool the surface rocks, but also attacked them chemically, breaking down the more soluble minerals. This released the more resistant crystalline debris to be washed into hollows in the rock surface where the pebbles, sands and clays from those weathered rocks began to accumulate. Chemicals released from the waters became deposited on the grain surfaces, often cementing them together, thereby creating the first sedimentary rocks. As time advanced, the surficial skin of solid crust grew, extending from the primeval cratons of the continental interiors to cover large areas of the globe but, as today, major hollows remained. These gradually filled with water to form the oceans, the floors of which have since become lined with sediment and organic remains. Thereafter the physical and chemical processes with which we are familiar today are believed to have been active on the Earth's surface.

Geophysical evidence shows that today the crustal rocks of the continents vary from 30km to 70km, averaging 45km in thickness, the greatest thicknesses being below mountain belts. By contrast, the heavier rocks forming the oceanic crust average only 8km in thickness. Despite the passage of enormous periods of time, the interior of the planet is believed to have changed little since its formation. However, the rate of escape of geothermal heat energy from the sub-surface has certainly decreased. Nevertheless, active centres of such heat losses may still be recognized in the volcanoes, from which hot lavas, ashes and huge volumes of gases, including steam, continue to escape to the atmosphere in many parts of the globe.

Most of the active volcanoes of the world are situated along lines following the ridges near the centres of ocean basins, as island arcs a short way offshore from the continental margins, or short distances to the landward of, and parallel to, the coastline within the continents. The Pacific Ocean is often said to be surrounded by a 'ring of fire', referring to its many marginal volcanoes, each of which marks a site beneath which the hot fluids within the lithosphere are injected upwards through the base of the crust. Fresh igneous rocks are added along the margins of splits opened through the mid-oceanic crust. At depth the circulation cell continues to move, carrying the newly created rocks away from the centre of thermal upwelling. These new rocks enshrine the magnetic characteristics of the time they were formed.

The Earth's magnetic field is known to reverse in direction from time to time, with the north and south magnetic poles exchanging dominance at irregular, but geologically recognizable, time intervals. On the ocean floor on either side of the crustal split that is marked by the central valley within the oceanic ridge where the new igneous rocks are formed, the patterns of the magnetic field changes are mirror images. This confirms that new molten rocks are repeatedly added into the crust along this central valley as its margins are slowly pulled apart by the circulating lithospheric currents below the thin crust. Iceland, formed almost exclusively from volcanic rocks, occupies a position astride the North Atlantic ridge, the oldest rocks of that country being on the eastern and western coastal extremities. The youngest rocks lie along the central zone of presently active volcanoes. Many other volcanoes occur in continental rift valleys, along lines or zones where the continental crust is being pulled apart by underlying migrating lithosphere, as beneath the East African Rift Valley.

The corollary of this process is that the oceans that have central, or near-central ridges, are becoming wider through time as the fringing continents are progressively swept further apart. Today the North Atlantic is said to be opening at about the same rate as the growth of human thumbnails. The oceans cannot continue to grow indefinitely, and where the margin of the expanding ocean meets a similarly mobile sheet of ocean floor travelling in a different direction, powered by a separate convection cell, there is a collision and the ocean floor of one system sinks beneath that of the other system, usually generating earthquakes. Each separate mobile mass of oceanic crust is referred to as a plate. Plates composed of continental material are thicker but lighter than those of oceanic plates, and so collisions between the two plate types usually lead to the oceanic plate gradually sinking back into the lithosphere beneath the margin of the continent. This is the process of subduction, of which the western coastline of South America is a good example with a zone of earthquakes and volcanoes landward from the coast. Most collisions of the plates are oblique to the continental margins rather than direct head-on events, and these give rise to fracture planes, faults, along which the crustal blocks travel past each other. The friction between the adjacent blocks gives rise to folded rock sequences and, very importantly, produces systems of parallel faults, as in the well-known San Andreas and Hayward fault system in California and in Scotland's Midland Valley, with many subsidiary faults parallel to the coastlines of the Firth of Forth.

Chapter 3

Scotland rocks, bends, breaks and migrates

The earliest rocks that we know in Scotland are confined to the North-West Highlands and the Hebrides. They lie within the Lewisian zone of highly metamorphosed (physically altered and at least partly re-crystallized) sedimentary and intrusive igneous rocks, which have experienced several periods of mountain-building and destruction since they originated between 3.1 and 2.7 thousand million years ago. In the process, the rocks became intensely heated and deformed during each construction phase. The Lewisian rocks have their direct equivalents in Labrador where, between 1200 and 1000 million years ago, the Grenvillian mountain-building event took place towards the margins of the early super-continental mass known as Rodinia.

As these mountains were uplifted, the rocks became exposed to the processes of weathering and erosion. We believe that these processes, which we see operating today, progressed in much the same way throughout geological time, and this enables analysis of the destruction products, which are now the sedimentary rocks, to reveal the conditions to which the Earth's surface has been exposed over the millennia. The aphorism *the present is the key to the past* has guided much of geological thinking over the last two centuries. Thus as the mountains were uplifted, rivers carrying the run-off from rainwater carved valleys into the metamorphic basement rocks. The destruction products of the erosion from these very early mountains are today seen in Scotland in the red, often pebbly, sandstones of the Torridon hills.

The Moine rocks, which occur mainly north of the Great Glen, are also principally of metamorphosed sediments from the younger mountain-building episode that originated between 1000 and 870 million years ago. South of the Great Glen, the Grampians are formed largely of Dalradian rocks, a third series of metamorphosed sediments and igneous intrusions formed between 800 and 520 million years ago at the edge of the Rodinian continental mass. South of the Highland Boundary Fault zone, the Dalradian rocks are believed to extend beneath much of the Midland Valley, but their southern limit at depth is not known at present (Bluck, 2002 in Trewin, 2002).

Below the continental crust of this northern area a subduction zone was developing around 470 million years ago, with the oceanic crust migrating northwards from the southwest and sinking below the over-riding continental crust. As the crustal floor of the expanding Iapetus Ocean pressed to continue its now obstructed northward movement, the downward travelling oceanic crustal plate served to partly compress and lift much of the outer part of the continental plate material. The sinking oceanic crust became reheated and the lowermost parts of the adjacent continental rock mass became incorporated in the resultant melt. Once sufficiently mobile, the melt became injected into the crust, sometimes emerging at the surface to give rise to a belt of volcanoes producing both ashes and lavas. In Scotland the remains of this belt of now long-extinct andesitic volcanoes extended along the line of the Sidlaw and Ochil Hills. Further north, more magmas intruded into the rocks of many parts of the Grampians to mark a more mature zone of melt production at still greater depth where slow cooling allowed the formation of relatively large crystals in the melts, giving rise to the granites.

Offshore from the continent, a series of volcanoes developed, possibly as an island arc in the north of the Southern Uplands, and together with material eroded from the surface of the northern continent, a wedge of sediments built southwards onto the approaching Iapetus Ocean floor. The sediments of the continental shelf, its margin and also the nearshore ocean floor became compressed and closely faulted. Today evidence of the compression is seen in the form of many northward-sloping faults, the fracture planes all indicating the general northward progress of the underlying sediments responding to the friction as the ocean floor continued along its downward path beneath the continental margin.

Beyond the line of the Solway Firth, the floor of the Iapetus Ocean continued to move northward or north-eastwards, carrying with it the Lake District, the rest of England and Wales, much of southern Ireland and part of Newfoundland, all of which had previously formed part of a micro-continent situated on the north-western margins of the ancient African continental plate until around 480 million years ago. The precise timing of the collision between the continental blocks is not well defined. It is thought to have extended over several millions of years, most probably during the period 420–410 million years ago. In geological terms this time period spanned the boundary between the older Silurian and the younger Devonian periods (Trewin and Thirlwall, 2002). The Devonian sediments are best known in Scotland under the name 'Old Red Sandstone' made popular by Hugh Miller, the Cromarty stone mason.

During the early part of Devonian times land surface adjustments were taking place, giving rise to several phases of mountain-building. As the mountains were uplifted the processes of erosion began to strip away rock materials to deposit them as sands and gravels in basins between the upland areas, gradually burying the old landscapes. Associated with the subduction and melting of the rocks at depth, a zone of volcanic activity developed within the north-eastern parts of the Midland Valley, yielding andesitic lavas and volcanic ashes that dominate the northern coastline of Fife. Between the lava flows there are coarse gravelly deposits (conglomerates) indicating that volcanic action was discontinuous and that the river-dominated erosion processes continued during the intervening quiescent periods. A second phase of mountain-building during the Middle Devonian led to the end of the volcanicity and allowed renewed uplift and erosion of the earlier deposits. In the east of Fife there was a considerable time gap, possibly as much as 10 million years, between the cessation of volcanic activity and the earliest succeeding sand and sediments, some of which were water-lain in rivers and coastal environments, while others formed wind-blown dunes.

The migration of major plates of oceanic or continental materials in relation to one another must also raise the question of where on the Earth's surface all this movement was occurring. Measurement of the Earth's present magnetic field has already been invoked to indicate the mirror-image pattern of changes in the relative dominance of the north and south magnetic poles in the rocks of the ocean floors. Today the global magnetic field is approximately vertical at the poles and horizontal around the equator, with intermediate angles of declination between these extremes. If this simple pattern existed in the past, then it is possible to determine at least the latitude at which the various continental plates were situated by measuring the ancient magnetic fields preserved in them. Initially, this was achieved using igneous rocks, which normally have iron-rich minerals crystallized within them and so preserve evidence of the magnetic conditions under which the rocks solidified.

At the start of Devonian times Scotland was about 30° south of the equator, and by the start of the succeeding Carboniferous period, it had migrated further north to around 15–20° south. In terms of the modern distribution of climate belts this represents a change from arid or semi-arid conditions to sub-tropical grasslands, and as the Carboniferous continued, the northward migration carried the area ever deeper into the tropical rainforest belt until, by the middle of the Lower Carboniferous (335 million years ago), Scotland was situated close to the equator (Read *et al.*, 2002 in Trewin, 2002). By the end of the Carboniferous (305 million years ago), the movement had carried the area to the northern limits of the rainforest belt and the periodic influxes of red sandstones typified the arrival of the northern tropical deserts of the Permian period. By this time the main continental plates are known to have come together to create the massive American, African, European, Asian and Australasian continent known as Pangaea, perhaps the last time that the major lands of the world have all been together as a single unit.

Chapter 4

Life emerges from the seas and invades the land surfaces

There are many theories and beliefs concerning the origin of life on Earth. Suffice to say here that simple life forms, initially single-celled organisms, are thought to have given rise to increasingly complex multicellular organisms through time. These probably lived in water, whether on land or in the seas, wherever they found sufficient nutrients to satisfy their needs. As these creatures were without exception soft-bodied, the chances of their becoming preserved in what we see as recognizable forms were vanishingly remote. It was not until many millions of years later, somewhere between 600 and 545 million years ago, that the organisms developed the ability to secrete sufficiently robust skeletal material that there was any opportunity for their remains to become preserved in the rock record, and then mainly in very fine-grained muddy sands. Now known from late Precambrian rocks of most continents, but first recognized in Namibia, the often very exotic-looking creatures of the Ediacaran fauna have different characteristics in Australian rocks from those of America, and those differ from the Russian and Asian forms. However, none have as yet been recognized in Scotland. Prolific life forms developed widely after the dawn of the Cambrian period 545 million years ago, spreading in all oceans, the hardest parts of the skeletons serving to generate limestones in many areas, as in the Durness Limestones of the northwest Highlands.

By the start of the Devonian period fish of many forms, most with external scaly or bony armour, were to be found in the seas around the Scottish coasts. Some of the marine creatures, notably the plants and fishes, were finding their way to survive in the brackish waters of estuaries and even in the lower reaches of rivers, where lack of familiar salts in the sea waters presented considerable challenges to their survival. The nourishment for the fishes would have been from planktonic organisms, burrowing worm-like creatures, and from plants grown on the banks. The plants were almost certainly the first organisms to emerge from the water on to the land surface, initially through becoming periodically stranded by the falling waters during the twice-daily retreat of the tides. Some of these plants were already able to make use of the solar radiation, as they and many of their forebears had lived largely in shallow waters readily penetrated by sunlight. They later became adapted to live largely free from the tidal waters and thrive on land, albeit necessarily remaining near a reliable source of water. Those best adapted to the new circumstances were what we know today as C3 plants, able to extract carbon directly from the atmospheric carbon dioxide. Such plants ideally thrive in hot, damp environments where there is high intensity light, typically in tropical conditions. A by-product of the metabolic processes that fixed the carbon was the release of oxygen to the atmosphere.

Preservation of the remains of the early land plants is very largely dependent on the nature of the environment in which they lived. Most commonly the plant remains preserved in the early Devonian rocks are mere traces of the small, fragile life forms. In many cases they were buried in sandy flood deposits. These sediments typically drained quickly, and as the organic matter degenerated rapidly due to oxidation soon thereafter, its destruction ensured that little fossil evidence remained. Later in the Devonian tougher tissues were developed, and the traces of these more robust plants were entrapped in the flood deposits to become preserved as fossils where the sediments were not quite so rapidly drained. The remains of the plants *Parka* and *Zosterophyllum* may be found in the sandstones of the Dundee area, along the Tay Estuary shoreline, in the Howe of Fife and parts of the Lomond Hills succession of rocks.

As we have already seen, throughout the Devonian period the Fife area had been carried slowly northwards towards what are today the climatic zones of the savannah grasslands and tropical rainforests. At the start of the Carboniferous the wetter climate allowed the development of powerful streams that carried sands along channels leading from a north-easterly direction across the eastern parts of Fife (Greensmith, 1965). The warmth and very ample supplies of water encouraged the plants to proliferate in form and in their density on the ground. From the earliest parts of the Carboniferous, colonization of the land began in earnest with growth mainly in the lowlands of the river flood plains. Although most prolific and varied near the valley floors, the hardier plants soon spread to cover much of the land surface. The plant tissues were principally composed of carbohydrates such as cellulose, lignum, starches and sugars, with lesser quantities of the resins, oils and waxes, which protected the more delicate parts of the plants from decay. The decay and decomposition to give rise to peats was principally brought about by the activity of micro-organisms in processes of bacterial fermentation. The degradation products were largely aliphatic acids, humic acids, aldehydes and alcohols. Any gases evolved during the processes are thought to have escaped into the atmosphere. Such coals are the humic coals. A second pathway of coal preservation is that of putrefaction in stagnant swamp water that is depleted in atmospheric gases and forms 'sapropel', an organic-rich mud that dries to a hard, tough and largely structureless form of finely textured muddy coal.

Many of the newly developed genera of plants such as the Lycopsids *Lepidodendron* and *Sigillaria* and the horsetail, *Calamites*, grew to considerable heights, with trunks stretching 40m and 20m respectively from the forest floor. *Lepidodendron* was a most imposing tree, its trunk bearing a scaly pattern and rhomboid leaf scars (Figure 4.1). At its base the trunk was often 50cm in diameter. Below ground its root system was essentially horizontal and its many bifurcations provided stability for the slender, non-segmented trunk, which was devoid of branches and reached a culmination of protruding club-like reproductive structures known as *Lepidostrobus*, each of which was up to 30cm in length. From these cone-like structures, spores were released into the air to be spread by the wind. In the absence of branches, groves of the slender, isolated trunks grew in the environment fully open to the skies.

The trunks of the horsetail *Calamites* are known to have been up to 60cm in diameter (Figure 4.2). They gave rise to many branches, all of which, like the main trunk, appear to have been sub-divided into units not dissimilar from the modern reed *Phragmites*, with rosettes or whorls of leaves growing from the nodes. Between the nodes the trunk was formed of longitudinally arranged woody tissues that projected radially into the central cavity of the trunk. Sporangial cones were located at the tips of the many branches.

However, not all the plants were giant trees. Dense stands of the much smaller, genuine fern-like plants such as *Pecopteris* (Figure 4.3A), and slightly younger and more advanced seed-ferns *Alethopteris* and *Neuropteris* (Figure 4.3B) formed a substantial undergrowth. The larger tree ferns, with trunks rising to more than 7m, had individual leaf stems up to 3m in length and basal trunk thicknesses of over 1m. Thick development of root systems was normal. The *Pteridosperms,* or seed ferns, were now long-extinct organisms that flourished in the late Carboniferous.

The debris of dead leaves, leaflet fronds and spores, together with rotten trunks of fallen trees, commonly hosting fungal colonies, contributed to a mass of organic debris that gave rise to a very productive layer of humus from the forest floor. The humic material would have held a lot of moisture, which in the tropical forests of today falls almost continuously from the tree-top canopy. This water would have prevented the oxidation of the plant remains, which had been largely responsible for the destruction of plant fragments during the preceding Devonian, and replaced this with reduction processes, which ensured the long-term survival of the remains. In many ways such an environment is ideally suited to the requirements of detritivore organisms, insects, spiders, scorpions and millipedes. Millipedes such as *Melanopsis* left many tracks in the humic sediments, suggesting that they were real giants, some believed to be over 3m in length. Their smaller cousins and the early forms of dragonflies also proliferated, and the remains of early amphibians have been recognized from some of these deposits.

Figure 4.1 Pattern of leaf scars on the surface of Lepidodendron stem (red-handled penknife measures 19cm).

Figure 4.2 Outer surface of two Calamites stems (scale in centimetres).

Figure 4.3A Leaflets of *Pecopteris* attached directly to stem (scale in centimetres).

Figure 4.3B Leaflets of *Neuropteris* attached to secondary twigs (scale in centimetres).

Beneath the humus, the soils were penetrated by roots from the plants, and often these soils can be seen to have been depleted in goodness, becoming pale-coloured clays or sands. These seatearths, which include the fireclays, were formerly worked in many places as largely inert raw materials for the ceramic industry. Above the fireclays, the humic materials sometimes built to considerable thicknesses.

Chapter 5

Coal formation

The humic soils or peats, formed principally of plant material, gradually dried out and were buried beneath flood deposits of sand or mud. As they became buried, so the waters and volatile gases were gradually squeezed from the peats, a process that effectively allowed the percentage of their carbon content to increase to form lignite, the basis of the much younger brown coals of northern Europe. As the burial continued, the overburden pressures continued to rise, and the peats were transformed into black sub-bituminous, later into bituminous coals with decreasing content of volatiles, and finally into the very valuable anthracite. The succession from peat to anthracite enables the coals produced to be referred to as 'having rank'. The highest rank coals are the anthracites and the lowest the lignites. Their internal layered structure is normally well preserved, with dull coals of highly degraded plant debris, or bright layers of the more woody original material readily discernible. It has been variously estimated that for every 1m of coal thickness seen today, the original peat from which it was generated would have been between 5 and 20 times that thickness. Some of the coals in East Fife are 4m thick, so that the original humic soil thickness would have been between 20m and 80m. The latter would have taken a long period of stable conditions to accumulate. Most of the coal seams in Fife are much thinner than this figure, many being no more than a few centimetres in thickness today.

Once a peat layer became buried beneath overlying sands or muds, the chemical or biochemical decay slowed and eventually ceased completely. As the thickness of overlying material increased, so the peat became compressed and the temperature of the surrounding sediments increased with the burial. Any remanent water and other volatile materials were progressively driven out of the deposit, and the peat became increasingly solid and enriched in carbon.

The humic coals, familiar in domestic use in the past, often preserve a distinct layered structure with 'bright' or 'dull' appearance according to the material from which they were generated. The coal types of clarain and durain dominate the 'bright' coals. The first of these is of highly lustrous layers of intensely black vitrinite, formed from bark and woody tissues that are believed to have collected carbon-rich fluids during their decomposition. The 'bright' coals also contain thin layers of micrinite and fusinite, believed to have originated from fine organic dusts under water or exposed to the atmosphere, respectively. When the peat was flooded the harder and more strongly banded coal of durain was generated. Almost entirely of micrinite and fusinite, the durain is an intermediate form between the humic coals and the sapropelic coals.

The sapropelic coals are usually brown-black in colour and of greasy appearance. These hard coals occur in relatively small, isolated basins within coalfields and comprise myriads of small shards of small scraps of plants with very little micrinite or fusinite (Adams, 1960). Their importance is largely related to the relatively high content of oils and gases. As such, they were much sought after throughout Fife and the Lothians in the nineteenth century.

One of the outstanding characteristics of the sedimentary rocks of Carboniferous age is that in most of the coalfields in Europe and North America, the rocks show regularly repeated successions of limestones, followed by mudrocks, which become more coarsely grained upwards. These are in turn overlain by sandstones, which pass upwards into ancient soils, often bearing traces of plant roots, above which are the coal seams. Above the coal seams there is normally a return to marine clays or limestone marking the start of the next cycle of deposition. This form of repetition is noted in the coalfields of the East Neuk and reappears in many of the coal sequences of slightly younger age in Central and West Fife. Referred to as 'cyclothems', they reflect repeated changes in the depositional environments in which the economically valuable materials accumulated.

The search for a mechanism whereby such repetition of depositional material could occur on a geographically large scale led to a concentration of research in deltaic areas, for it was known that large deltas with their migrating channels carrying heavy loads of sediment were areas of rapid changes of depositional environments. The Scores Cliff (Witch Lake) at St Andrews provides good examples of deltaic sedimentation not yet developed into full cyclothems (Figure 5.1).

Repeated forward growth of deltas being built by rivers in front of prograding coastlines would allow marine limestones offshore to become progressively buried by the advancing delta front, bringing muds, silts and later sands to the area as the deltas extended seaward. The forest plants were then believed to have colonized the exposed tidal beaches and sand flats. It was suggested that, as the underlying sediments lost their water content and became compressed by overlying deposited sediments, space would have been created for renewed late phases of delta building. Within adjacent deltas such patterns might be recognized and correlated, but in coastal areas into which little sediment was carried, there could be only very thin deposits with little confidence in recognizing continuity of the deposits. This form of model might satisfy conditions of a single delta or set of neighbouring deltas as they evolved, but it did not

Figure 5.1 Witch Lake Cliff, The Scores, St Andrews at low tide. Cliffs mainly of sandstone dipping beneath the town with three coal seams.

satisfy the fact that in many cases individual coal seams, marine sediments and entire cyclothems may be traced unbroken for very large, even continental-scale distances along their outcrop. Not all rivers build deltas and some do so much more rapidly than their neighbours. Inevitably, substantial lateral discontinuities in the sedimentation patterns should occur between the adjacent river outfalls from greatly contrasting catchments. So while the model concept was sound in itself, it did not satisfy all of the criteria required. In particular, it was largely constrained by the assumption that sea level would be constant or at least very stable through time.

A mechanism allowing for the repeated changes of sea level on a global scale was lacking until it was recognized from seismic stratigraphic analysis in the hydrocarbons industry that worldwide changes of sea level by 30m to 100m could be recognized in the geological record (Vail and Mitchum, 1977). In particular, such variation could be achieved during periods of glaciation, with growth and destruction of major ice sheets, and furthermore that such changes had been repeated many times over a substantial span of geological time (Van Wagoner *et al.*, 1988).

A geologically recent model for this form of sedimentation pattern is known from the recent Ice Age, when there was repeated growth and melting of glaciers and major ice sheets. Water was removed from the seas and oceans leading to widespread falls of sea level in order to provide sufficient water for the ice sheets to form. As the waters drained away sea level fell, to be followed by matching rises in water levels once the ice sheets melted.

It is suggested that in the early part of the Carboniferous repeated episodes of withdrawal of the sea allowed soils to develop and the coastal forests to expand seawards and colonize the newly exposed areas of former sea floor. In time the rivers in the tropical regions built forward, laying down fresh alluvial sediments that buried the early-established forests, creating non-marine deposits where there had previously been only fully marine depositional conditions on the sea floor. Onto this newly lowered 'land' surface, the forests swiftly extended and thrived until the melting of the ice sheets in the high latitudes around the poles returned their waters to the oceans, allowing sea level to return to its 'normal' pre-glacial or more correctly, inter-glacial position.

As the sea flooded in, the recently colonizing forests became drowned and their humic soils became buried beneath sediment from the eroding coast. In place of the coastal forests, now delta slopes, coastal lagoons and tidal flats were formed. On the beds of some of the lagoons and the adjacent sea floors shelly organisms, crinoids, corals and shellfish produced calcium carbonate skeletal remains that subsequently gave rise to layers of limestone. Although in central Scotland the limestones are normally less than 2m in thickness, some are over 5m thick. Elsewhere the lagoons became filled with clays, now converted to shales often containing the remains of planktonic plants and sometimes bearing oil and gas. The shales commonly grade upwards into fine-grained, often rippled sands bearing marks of having been accumulated in tidal conditions.

Is there any evidence to support the idea of glacial activity away from the tropics at that time? While the earliest, truly ancient settlers in Fife were enjoying the warmth of the sub-tropical or tropical sun, the ancient continental mass extended much further southwards across the desert-like and temperate climatic zones, and in what is today Southern Africa, with its then clustered neighbours of South America, Australia and the Indian sub-continent, conditions were very strongly circumpolar. For well over a century, evidence of glacial activity during the Carboniferous has been known in these countries. Ice sheets of the Carboniferous had left clear marks of their occupancy of the area in the form of scratches or striae on bare rock surfaces and in the remains of the glacial tills that covered the older rocks exposed at the land surface at that time. On the evidence of the recent Ice Age, the rise and fall of sea level is unlikely to have been a single event but would have been repeated several times, with patterns of change of offshore sediments reflecting each phase. There is no reason to suspect that the advances and retreats of the ice sheets during the Carboniferous would not have taken place many times.

The impacts of the repeated rises and falls of sea level were simultaneous across the world. In Fife there is evidence of at least 46 periods of rise and fall of sea levels, giving rise to substantial deposits of limestone or marine clays in the early part of the Carboniferous between 342 and 316 million years ago in the Tournaisian, Viséan and Namurian stages (Table 1). Similar conditions probably recurred during the Westphalian, the last phase of the Carboniferous. No deposits of that age are found in East Fife, but they gave rise to very significant reserves of coal in the Dysart to Leven and Markinch areas to the west and also in the Westfield basin near Lochgelly. During the earlier part of the Carboniferous, most of England and Wales had been covered by open and quite deep marine waters, creating thick limestones, and so did not benefit from the repeated land emergence and flooding events that led to the formation of the early coals in Scotland. Virtually all of the important coal deposits south of the border accumulated during the later phases of sedimentation. In the east of Fife many thin, but few thick, coal seams were developed during the earliest parts of the Carboniferous period.

Table 1: The principal stratigraphic names applied to the rock units.

1	2	3	4	5
320	Namurian			Castlecary Limestone
		Arnsbergian	Upper Limestone Formation	Index Limestone
			Limestone Coal Formation	Top Hosie Limestone
326.5	Visean		Lower Limestone Formation	St Monance Brecciated Limestone
331		Brigantian	Pathhead Formation	
		Asbian	Sandy Craig Formation	
			Pittenweem Formation	
			Anstruther Formation	
342.5		Holkerian & Arundian	Fife Ness Formation	

Column 1 gives the radiometric age of the base of the unit in millions of years (Ma); **Column 2** lists the main rock series within the Carboniferous; the Viséan extended to the top of the Lower Limestone Formation and the Namurian, thereafter to the top of the Passage Formation; **Column 3** gives the stage names; **Column 4** lists the Formation names known in the east of Fife; **Column 5** gives the name of the basal indicator Limestone of each of the younger divisions. (After Trewin, 2002.)

Chapter 6

Historical framework for mining in Fife

The earliest record pointing to early man having knowledge of coals is said to have been the finding of a flint axe head in a coal seam at Craig Y Parc in Monmouthshire, (Fordyce, 1860). During the Roman occupation of Britain, ashes from burned coals have been reported from several settlements. At Benwell (*Condercum*), on Hadrian's Wall the smithy in the workshop of the third fort from the North Sea was heated from locally supplied coals in the second century AD (Graham, 1979). There is also evidence that coals were used for cremations near the western end of the Antonine Wall in the Forth–Clyde Valley area during the mid-second century AD, (Montgomery, 1994). In the Tungian settlement at Vindolanda 120–130 AD, a jet toggle and fragments of high quality coals were recovered from a Period V building, one of many from which coal dust was found, suggesting that the coals were not used exclusively for metal working at that time. There is no direct record of coal use at the late second century to early third century Severan fort at Carpow, (*Horrea Classis*), on the Tay estuary near Newburgh, which lay well to the north of the Fife coalfields. According to the archaeologist and historian J.J. Robertson (pers. comm.), the historian Tacitus recorded that the Roman Empress Julia Domna entertained the troops at Carpow, which functioned as a forward base for legions preparing to strike north. It was supplied by sea from the Tyne. In addition to carrying the troops and their military hardware, the cargoes could have included coals for the use of the artificers and for heating the hypocausts in the main buildings of the camp.

Bede, writing in northern England in the seventh century, commented that coals were used as items of merchandise, but there is some dispute over translation as to whether he really referred to the jet worked as trinkets at that time. The Saxons referred to the material as *Groefan*. From the five or more pre-Norman centuries, when the Picts occupied the eastern half of Fife, including the Abernethy area adjacent to Carpow, as part of the Kingdom of Alba, there remain no written accounts of the use of coal in the area, although intricate metal-working from that period is known.

During much of the Dark Ages, the indigenous population became largely divided along tribal lines, with families showing allegiance to a particular leader or locality, where they worked the land as farmers subject to traditional and often unwritten legal and administrative constraints. The peasants, who lived in wooden-walled thatch- or turf-roofed huts, cooked in peat- and wood-fired cooking pits that were situated on the floors of their dwellings. The smoke normally circulated within the property, as there was rarely more than a hole in the roof or a doorway through which it could be cleared from the living and sleeping quarters.

The arrival of the Normans in Britain led to an attempt to establish uniform customs across the island. The king granted land to his nobles or lords, who then assumed responsibility for the people living on those lands. Each lord was the king's vassal and was granted a 'fief', for which the vassal was expected to undertake some special service for the king. He was normally required to build and maintain a castle to help keep order among his people, to make some tax payments, and in addition, he was commonly required to appear armed and on horseback as might be needed by the king. The vassal could sublet part of his fief to local prominent people, the mormaers, who could in turn let to peasants who had no rights beyond being allowed to work with their hands, cultivating the land and paying taxes or feus in return for the lord's protection. If the peasant failed to pay his feu, the land could revert to the lord, and if the lord failed to complete his fief, the lord himself would be in similar trouble from the king. This feudal system was already well established in France at that time. The lord's stone-built castle was normally provided with chimneys leading upwards from fireplaces where wood or peat was burned. Whereas the lord could keep his fire burning at night, the peasants were required to extinguish their lights and fires in response to a curfew bell (the *couvre feu*).

Britain had become fully feudalized by the end of the reign of William the Lion (1165–1214). The Scots King Malcolm Canmore and Queen Margaret were francophile in their outlook, and their son King David I (1124–1153) had spent 40 years at the Norman-dominated English court before assuming the Scottish throne. It was natural for him to perpetuate the feudal arrangements.

King David favoured the Cistercian and Tironensian monasteries with their reverence for hard work and withdrawal from the world. The world of mining was well suited to the ecclesiastical establishments. In 1202 the monks of Holyrood were granted a tithe of the coals from Carriden and in 1210 and 1219 the Cistercians at Newbattle Abbey received the rights to work the Tranent coals, possibly in opencast workings (McKechnie and Macgregor, 1958). Until this time all the coals collected and used domestically were referred to as 'sea-coal', as any coals used in coastal communities were sourced mainly from the shorelines or river bank exposures. The bulk of the population used peat or turf for heating. The earliest known reference to coal in the east of Fife is believed to be from a charter of 1235 when a boundary description at Caiplie (Kilrenny), stated 'so eastwards along the shore as far as the white skerry beyond the colepot' (coal pit?) (Taylor, 1903, Vol. III p.329).

The earliest known charter in Northern England was issued by King Henry II to the townsmen of Newcastle in 1239, granting permission to work the coals in both the Castlefield and the Forth.

The Cistercians developed special expertise in mining for both coal and lead, and a charter of 1291 allowed the monks of the Benedictine Abbey at Dunfermline to take coals from the nearby Pittencrieff Estate and Calurig areas strictly for their own use, and in 1294 the monks of Paisley Abbey were permitted to work their nearby coals (Figure 6.1). At some localities the colliers were said to work in such dreadful conditions that the mines were commonly worked by chained prisoners as part of their punishment for wrong-doing.

By the start of the fourteenth century coals were being carried to London from the coastal mines in northern England and central Scotland. Whilst the coal fires were entirely suitable for heating large stone-built buildings such as castles, monasteries and many manor houses, most of the buildings used for the dwellings of the less fortunate members of the population remained essentially unchanged, with wooden walls. Heated from stone-based cooking hearths, often centrally placed within one room, few had stone routeways to lead the resulting sparks and smoke from the premises, and lacking formal chimneys, any smoke escaped through the thatched roofs. Awareness that living in noxious, smoke-filled houses was deleterious to the health of the populace triggered a Royal Proclamation in 1306, which banned the burning of coals in houses and the use of metal-working furnaces in London. While this represented a considerable set-back to the coal industry, the problem of domestic heating was overcome by the successful introduction of the fire-grate and an iron-made chimney structure which could be moved to different positions in the building as needed (Galloway, 1882).

Figure 6.1 Medieval Cistercian monks with their tools. [Cambridge University Library MM.mm.5.31 fol.113.]

The establishing of 'royal burghs', 'episcopal burghs' and lesser 'burghs' of descending order of importance was designed to stimulate trade. These burghs functioned as administrative centres, collecting taxes and housing the major and smaller law courts according to size. The larger units were responsible for substantial areas. For example, in Fife, the Cupar liberty extended along the Tay estuary from Balmerino to Newburgh, south to Leven and Largo, but had restricted access east of a line from the Tay to Dura Den and Peat Inn. The royal burghs had exclusive rights for the export of goods, and this precluded export to foreign countries from many of the coal-producing areas. The administration of relatively minor court cases was dealt with by the barony courts and only the more substantial cases were retained for the attention of the sheriffs.

There was no long tradition of servility or slavery in Scotland through the taking of prisoners in military actions, as had been common in Europe from Roman times onwards. However, following the Norman custom, in 1070 so many prisoners were taken during a Scottish invasion of northern England, that it was said that few settlements in Scotland lacked either slaves or handmaidens from that source. These were some of the earliest recorded serfs, the *nativi,* who could be bought or sold as goods, and as a result of which, many were deployed in the coal industry. Their distribution through the country was effectively the first step towards a fully feudal system, but for 200 years after the middle of the fourteenth century, few references are made to feudalism by historians.

The working of the coals and the associated limestones, principally for the production of mortars for major buildings, certainly continued. Many of the coal workings had extended as far from the surface as was technically possible at that time, with groundwaters presenting problems curtailing the deeper penetration in many areas. Although there are few documentary references to the coal industry during the fourteenth and early fifteenth centuries, it is known that the workings of the coastal salt pans, which relied on the heating from the local coals to drive off the water to concentrate the brines, continued throughout this period. At this time all of the salt provided for the populace was sea-derived. Now, more than six centuries later, the recent culinary fashion has returned to using the salt as used by the peasants of the distant past.

By the middle of the fourteenth century the coal production from quarries and coastal sections was being overtaken by the development of workings in vertically shafted bell pits and horizontally driven adits. These were very successful where they were able to work above the level of the water table, the upper level of the rocks permanently saturated by the groundwaters. Where drainage systems could be constructed, the workings could be continued, but in many cases problems with the drainage of these waters led to the closure of the workings.

The decline in recorded activity in the coal industry at this time cannot be explained entirely by the practical problems of the mine workings. The fourteenth century was a period during which illness repeatedly swept through Britain, as also on the continent of Europe. Before the arrival of the flea-infested rats carrying the Black Death in 1349, there had been a slow increase in the population, which was well able to work in the mines or in the fields for the production of food. However, it is estimated that, after the passage of three or more plague outbreaks before 1401, the population of Scotland had been reduced to around two-thirds of its previous strength (Smout, 1998). This is an almost identical estimate to the one-third reduction (by 100,000) in the population of Wales during the same period, as recorded by Davies (1993). Galloway (1882) reported that in 1351, and later in 1358, the burgesses of Newcastle successfully sought funds from Edward III for assistance to support the coal workings where the workforce had been severely reduced by visitations of the plague.

The passage of the Black Death was followed by other fatal epidemics in 1430 and 1450, with milder occurrences in 1500, 1514, 1530 and 1539. According to Creighton (1965) at least five more medical reverses before 1608 led to a further 20% loss of life across Scotland, with Leith losing nearly 50% of its population from typhus. Mild outbreaks of smallpox killed 1 in 7 of the population of Aberdeen and Glasgow, moderate attacks 1 in 5 and severe epidemics 1 in 3 or 4 between 1635 and 1672. Similar attacks accounted for 19% of all deaths in Glasgow between 1783 and 1800. In addition measles, typhus and diphtheria were killers in the early nineteenth century, while water-carried typhoid and diarrhoea were recognized as having further annihilated large numbers of the populace in the absence of the other diseases. There can be little doubt that the spread of these ailments was aided, if not caused, by the almost total lack of organized disposal of human and animal wastes. In 1568 Gilbert Keyne referred to 'the stinks of corruption from human ordure quhilkis occupies the commune *strettis* and *gaittis.*'

In Scotland, as a result, there were scarcely enough people to work the land; the production of food assumed great importance, and so few people were sent to toil below ground. The landowners were less interested in having slaves and serfs as colliers than in having them as paying tenants on the land. In the fourteenth century no Scotsman was a serf and the worst excesses of feudalism vanished for more than 200 years. Historian John Major (Hume Brown, 1893) noted that by1521, the peasant classes had largely become tenants renting land from the local laird to whom they owed allegiance. Although the mining industry was not completely shut down, its evolution undoubtedly suffered as a result of the diversion of labour to the fields.

Nevertheless, through time there was a growth in the use of bell pits to give access to the seams. These, and vertically shafted workings, extended deeper and deeper, but inevitably problems with groundwater were encountered in increasing numbers of sites and removal of the waters by pumping was recorded as early as 1486. Many of the coals released gases into the workings. Some of these were potentially explosive, and occasionally seams caught fire spontaneously. The Dysart Main Coal in the south of Central Fife burned below ground for over 200 years, giving the name of Hot Pot Wynd to a village street near the harbour. The Main Coal in the Drumcarro coalfield in northeast Fife is known to have burned below ground for more than 50 years in the mid-nineteenth century. In neither case did these fires interrupt workings in coal seams above or below.

The threats to availability and regular delivery of coals to the London area led Edward III to become interested in the means of transport, particularly from the Tyne area to the Thames. He sought to regulate the coal trade by controlling the quantities of coal assigned to individual vessels, and achieved this by favouring the keel, a broad, flat-bottomed craft with a shallow draft. This was able to travel upstream from the sea (Figure 6.2). The keel, already widely used in British coastal waters, carried a total of 20 chaldrons of coal, each of which would attract a tax of two shillings and two pence, (Duckham, 1970). The London chaldron contained 36 bushels or 25–27cwt, but the Forth chaldron was approximately 30cwt. For some years the keel, (about 25 tons), became a measure of a large quantity of coal as supplied to London and further afield.

Figure 6.2 A 'Keel' boat loading coals at the riverside. (Watercolour painting by J.C. Ibbetson, 1796.)

Each time the population began to regain its numerical strength and sufficient hands became available to supply both the agricultural and industrial needs, there was a renewal of workings in the temporarily abandoned coal heughs. However, by 1561, when Mary Queen of Scots returned home, the total annual Scottish coal output is estimated at approximately 40,000 tons, (Arnot, 1955). The resurgent industry first peaked in Scotland after 1580. The appropriately named 'Cromwell's Mine' which Dron (1902), reported to have been working before the Reformation, was later overseen by Thomas Tucker, appointed in 1655, on behalf of the Parliamentarians (Smout, 1998). It was a near-surface working on the southern margin of the Pittenweem coalfield. This field, destined to remain in production for a further two centuries, was part of the early resurgence. Demand for the export of coals to Europe increased greatly during the sixteenth century and a lucrative trade developed, in which coals exported at 2s 2d a chaldron from Tyneside sold for as much as £4 6s 8d in France. In 1546, Henry VIII ordered 3000 chaldrons to be delivered to his population in Boulogne. Of course, the export of coals to France led to shortages in the London area, to the extent that, in 1563, an act of Parliament was passed banning the export of coals from the country. At this time the supplies of wood available for domestic heating were dwindling and the introduction of coal as an alternative had been remarkably successful, so the ban was widely ignored.

The renewal of activity in the mines, which was limited by the numbers of people available to undertake the skilled tasks, provided the first steps in a return to the old employment conditions and made way for an eventual return of serfdom. In 1574, Queen Elizabeth's Parliament in London had abolished the condition of serfdom, but only five years later, the Scottish Parliament, recognizing problems of potential labour shortages, strengthened it by stipulating that females should be kept in bondage until the age of 18, but that the men should be so constrained until the age of 24 years. A second step recognizing the scarcity of skilled workers led to the passage of an employer-sponsored Scottish Parliamentary Bill in 1606 requiring that colliers or salters would need to provide letters of reference from their previous employer before they could move elsewhere for work. Failure to provide such a document meant that they could be reclaimed by their former master within a year and a day, taken back and punished for theft of themselves and the labour that they could have done during their absence. Arnot (1955), summarized the effect of the 1606 Act as 'succeeding in making the colliers into an enslaved class, degraded below the poorest of free men'.

Dott (1944) reported that, according to papers in the archives of the Durham Family, before Queen Elizabeth's successor, James I of England and VI of Scotland, moved to England, he used the Parrot coals from Fallfield [sic] west of Largoward for heating at Falkland Palace between 1590 and 1600, indicating that there was reliable systematic mining in that estate before the turn of the sixteenth century.

Within a generation, largely during the reign of Queen Elizabeth, coal replaced wood as the principal agent for heating and cooking in domestic premises in London (Howes, 1612 in Galloway, 1882). The coals were transported to the city by sea, mainly in relatively small vessels such as the keel, in scale very similar to the red-sailed Thames barges that worked into the twentieth century. There was concern that the supply by the Coal Fleet could be interrupted in the coastal waters of the North Sea by some unfriendly foreign power. As a defensive measure, a limited but armed naval protection screen was created. By the end of the sixteenth century the collier ships numbered 200 and, as demand for coal rose and the industry increased its output, the total fleet numbers increased to over 400 by 1615, half of which supplied London. By the 1630s between 600 and 700 were counted in the supply chain, and by 1690 there were over 900 involved. The transport of coals to London was interrupted during the Civil War and there were shortages between 1642 and 1644, and again during the war with Holland in 1667, but thereafter the trading was gradually re-established.

The second half of the sixteenth century was a period of upheaval with the demand for the re-organization of religion throughout Scotland following the arrival and rapid expansion of the Covenanting movement, which had become well established by 1590. The Reformers followed the concepts introduced earlier by Luther, Zwingli, Bullinger, Calvin, Knox and others in continental Europe, where there had been bitter disputes between Catholic adherents and the Reformers seeking flexibility within religion. In Scotland much of this conflict took place in the St Andrews area but rapidly spread across the country, to the extent that the Reformers were able to create an effective Scottish Parliament by the start of the 1640s. Almost inevitably, the Covenanters became involved in the English Civil War between non-conformist Parliamentarians, later led by Cromwell, and Royalists. Open warfare developed across Scotland, with the Duke of Montrose and many Highland clans supporting the Royalist, Catholic cause. The Presbyterian Covenanters demanded freedom of worship outside the structured format typifying the Catholic doctrines and their worship in Latin, which were remote if not inaccessible to the largely poorly educated ordinary people.

The Reformation took its toll on the populace of Fife as demands were put on the lowland burghs to provide foot soldiers to fight for the cause. Sibbald (1803) noted that in 1644–45 General William Baillie required no fewer than three regiments, each one thousand men strong, be raised by the traders, landowners, masters and crews of ships from all of the Fife towns and fishing villages. The prominent personnel duly sent representatives from their staff. As a result of poor training and dubious leadership at the Battle of Kilsyth, virtually all were slaughtered, with the bulk of more than one generation of Fife men wiped out by Montrose and his troops. Beyond the personal calamities for the families directly concerned, this represented a further huge economic setback for the entire area. The mining communities were not immune from these losses.

The status of the collier families received a further blow with the Scottish Act of 1660, which dictated that if a man agreed to work in a mine or a salt pan, he made a serf of himself, effectively becoming a part of the mine. Such a collier was counted *ascripti glebae* and, just like a piece of machinery, he could be sold on, loaned to others or inherited by the owner's heirs. Most colliers married other underground workers, and as wages were kept low, so that even with the provision of the commonly minimal housing and customary rations of oatmeal, the people were rarely able to afford the inevitable arrival of children. Many estates provided *arles*, initial funding towards the clothing and upkeep of children, but acceptance of this meagre assistance ensured that the child would be bound to the estate in serfdom for life, effectively having been sold into slavery. Few collier families could afford to refuse this aid, and so remained firmly trapped within the industry. The colliers' wives and daughters were not normally paid by the mine, their labour being seen as a natural part of the collier's work and included in his wages. In 1679 that pay would have been 7 to 8 shillings Scots (about 1 shilling English or 5p today) per week. If the collier did not do as the taxman or laird instructed, he could be placed in the jougs, an iron collar around the neck, and chained to the wall above or below ground, depending on the nature and place of the misdemeanour.

A major contrast between the Scottish and the often larger and longer-established mines of northeast England lay in the fact that serfdom of the colliers did not redevelop south of the border. The collier communities were therefore free from the demands to provide foot soldiers to satisfy the whims of the rulers of the day.

In practice, serfdom became increasingly common towards the end of the seventeenth century. It does not seem to have applied everywhere, and during the 1690s many colliers were able to move from mine to mine on one-year contracts. Nevertheless, the conditions in which the colliers worked continued to provide disincentives to potential recruits to the industry.

The mining communities were not alone in being dependent on the estates for their housing and employment, for the agricultural workers were subjected to a long-established system of bondage, very similar to that of serfdom. An agreement was made between the farmer and the unmarried son who had learned to work one horse with the plough. He became a hind when able to manage two horses. When he married he became formally part of the bondage system and moved from his parents' home to separate accommodation. His wife was required to work at the harvest and other labouring tasks as a bondager in return for the housing rent. When children were born into the bond, the family had to take in a further bondager to carry out the work formerly done by the wife. Typically, the bondager wore a long dress and a large wide-brimmed, high-crowned hat (Ibbetson, 1796 in Lewis, 1970; William Cobbett, ed. Ian Dyck, 2001) as protection against the sun when working in the fields. Although this bondage system existed in northeast England and southeast Scotland, there is no record of it having actually operated in the East Neuk.

After the Act of Union in 1707 the Scottish mines were allowed to export their coals to the colonies in North America. This gave a boost to the mines in the west of Scotland, but was less useful for mining enterprises on the east coast. As the Industrial Revolution gathered pace during the eighteenth century the demand for ever-increasing production of coal to provide the coke needed in the newly developed blast furnaces ensured that, in the face of labour shortages, inducements in the form of increased wages were once more needed to boost recruitment. The wages paid to colliers rose to 12 and even 13 shillings per day, which compared very favourably with those of agricultural labourers, who then received 4 to 6 shillings per day (Millar, 1771). In addition, the miners received a coal allowance, a free issue of carbolic soap and a supply of coarse (itchy) towels. While this led to a general improvement in the standard of living for workers throughout the industry, it also provided the opportunity for lairds to operate the truck system, providing over-priced food and general supplies sold to the collier families living at sites remote from villages and townships, to increase their own income. The new fixed wage structure did not find favour with Adam Smith (1776) who called for the introduction of financial incentives to encourage improvements in the efficiency of working.

There was a huge increase in industrialization throughout the eighteenth century. Competition for sales between the coal-supplying estates led to reductions in coal prices and a drop in the wages, linked to reductions in the number of days of working in the mines to three or four a week. Desertions increased as many colliers moved away to join the better-paid armed forces. Just how difficult were the living conditions? A short summary of the variations in wage levels given by Arnot (1955) is useful here to illustrate the difficulties faced by the colliers during a time of rising food prices:

Table 2: Wage variations for colliers 1812–1855.

Year	1812	1822	1841	1851	1853	1855
Average Weekly Wage	20/-	25/-	16/-	12/-	18/-	17/-

The average weekly wages, which had been at an acceptable level at the start of the nineteenth century, became severely eroded over the following 40 years, falling to 12/- in 1851 (the year of the Great Exhibition lauding industrial advances) before starting to recover to 18/- in 1853, but they were reduced again by 1855. The pattern of brief rises and sharp falls continued into the next century. Such drastic changes for families on or near the bread line brought considerable hardship across the collier communities.

The levels of deprivation and destitution experienced across the mining industry throughout the United Kingdom led to the miners becoming organized to seek the introduction of improvements not only in wage levels, but importantly, in the safety of working conditions below ground. Although Chartism as such decreased in influence in the 1830s, many of its aims were carried forward by the foundation of what were to become the Trades Unions following the establishment of the Miners' Association of Great Britain and Northern Ireland in 1841. By 1844 this organization began to seek an agreement with the mine owners that they should not reduce product prices and wages in competition with each other, but should agree to ensure that there would be sensible returns to each side of the industry. During this early part of the nineteenth century, in the face of constantly rising numbers of fatalities in the industry, the Miners' Association pressed very strongly for improved safety measures to be introduced, coupled with the introduction of formal education for collier children. The appointment of Inspectors for Mine Safety was introduced with support from colliery managers, mining engineers and coal owners from the newly formed Mining Institute of Scotland following the major accident at Blantyre in 1877, which killed 207 miners.

Those who became destitute through lack of jobs, old age or sickness required support from the parish, from societies in the community, or from the charity of the local coal proprietors and other worthy families. Parish support in Scotland was less than that given under the Poor Law in England and it was means tested in the extreme, so that very few qualified to receive even this support. Arnot (1955) quoted one parish in which 35 destitute families, totalling 185 people, received a total of 5/- per week between them from the Poor Law.

The story of the successful rise of organizations seeking to improve the lot of the colliers is one of considerable interest, and is thoroughly investigated in Arnot (1955), to which the reader may be encouraged to progress. While these aspects were and are still of undoubted historical importance, they lie beyond the remit of the present review.

Chapter 7

On-land transport of the coals

The road systems

In the earliest account of fuels being carried in Fife were those of Lindores Abbey at the start of the fourteenth century. Dowden (1903) referred to the movement of peats from mosses near Ladybank using oxen, horses and carts *(cum suis bobus et carris)*. The Charter of 1302 refers to the use of four-wheeled wagons *(boues)* normally drawn by oxen, and smaller two-wheeled horse-drawn carts *(plaustra)*. Such wagons commonly had solid wheels on a rotating axle.

For at least the next 400 years, the movement of the coals from the hillside heughs or bell-pit shafts was by means of pack horses with pannier baskets mounted on either side of the animals. The routes normally followed well-worn, long-established footpaths between settlements. Initially these led along relatively well-drained ridge tops, but much later alternative routes were developed towards the foot of the slopes still avoiding the valley floors, which were often poorly drained and subject to flooding. The low-level routes would have required the animals to cross water courses at fords or pack bridges, narrow structures rarely more than 2m in width (as in Kemback Bridge and Bishop's Bridge beside the Fife Folk Museum at Ceres). These largely unstructured tracks were used daily by the local people. According to Silver (1987) before the start of the sixteenth century, when a member of royalty or a senior administrator conducted tours of inspection, they required (but did not always find), the provision of much more formal routeways. A statute of 1285 decreed that the King's Highway was required to have a clear zone 60m wide on either side of the track when passing through woodland to ensure protection from attack. Such a route would certainly have been taken by the coal users of the day, should it have coincided with their needs. As late as 1555 maintenance of these major tracks was recognized as being of importance, and compulsory service from parishioners was required to provide any necessary repairs. The upkeep of the bridges was an act of piety or penance. Until the start of the seventeenth century virtually all overland transport was achieved by pack horse. The use of wheeled carts was negligible, although occasional records show that sledges were sometimes used for major items, especially during the cold winter of 1590 that marked an early stage in the oncoming Little Ice Age.

This system of essentially voluntary care of the roads was less than satisfactory. In 1610, when Justices of the Peace were introduced in Scotland, one of their main functions was to ensure the upkeep of the roads, requiring local workmen to be provided by the landowners. The normal roads were now to be 5m wide, with the major highways 6.4m wide and lined with gravel or stones to make possible the passage of carts and carriages. However, there was no sign of any significant overall road improvement in Fife before the 1633 coronation in Scone.

The coals from the south of Fife had been used for several centuries to power the beacon on the Isle of May, to provide the heating for the many salt pans along the coast, and for export to England and the Low Countries. The delivery of suitable coals to the ports required little transport along inland roads, rather than along routes linking adjacent coastal townships as in the south of Fife. An early account of an inland delivery of coals from the eastern mines was referred to by John Lamont (1810) of Newton (near Kennoway), who quoted farm diaries for the period 1649–72 recording provisioning of Falkland Palace with 36 loads of coals from Largo. In 1653 the coal mines of Falfield, west of Largoward, were ordered to provide a delivery of 36 loads of coals to troops stationed at Struthers Castle near Ceres.

During the mid-seventeenth century most of the farmed land became reorganized from rigs or strips of land with mixed and assorted tenancies to form larger coherent adjoining fields. The intensive farming of the rigs gave way to larger, more economic units than previously and the ultimate formation of the fermtouns with unfenced outfields reaching towards the adjacent hills with their common grazing lands. Fenced or walled infields lay within enclosures near clusters of small cottages where farming and mining communities lived. The agricultural production began to increase with land improvement through the addition of lime or seaweed to the soils and systematic drainage of the lower land. As crop production increased, the demand for road improvements increased so that the limestones and coals for burning could be carried to the fields. The result yielded improved quality and quantity of the crops taken to market. In 1669 and 1686 two Acts of Parliament appointed Commissioners of Supply to concentrate their activities on the production of bridges while the landowners took responsibility for the roads. One technique encouraged the farmers to gather stones from their fields and spread them on to the adjacent road surfaces.

It is known that in 1663 the major roads were of sufficient standard for Archbishop Sharp of St Andrews to travel by coach from Kinghorn to Falkland Palace, but for other journeys in the area, he reverted to horse and saddle. However, in 1679 he was in a coach approaching St Andrews when he was stopped and murdered on Magus Muir.

The end of the seventeenth century was a period of considerable hardship in rural areas, with a further advance of the Little Ice Age accompanied by substantial reduction in crop yields. Most of the heavy traffic on the roads was involved in the transport of limestone and coal with agricultural produce accounting for the rest. Short lengths of well-constructed roads in the south of Fife led from the mining areas such as Balbirnie and Rothes estates to the ports such as Dysart and Kirkcaldy, which handled the coal exports. The same two estates, owned by Lords St Clair and Rothes, combined their efforts in 1740 to construct a good quality road north across the Riggins towards the lime kilns at Forthar, and later on towards Newburgh and Cupar.

Following the Jacobite Rising of 1715, Major-General Wade was appointed to construct military roads between the major forts in the Highlands. His design featured a basal ditch 16ft (5m) in width, lined with fitted stones and covered with 3ft (1m) of gravel. On either side the materials from the trench were used to form low ramparts. Between the roads and the ramparts, many landowners left shallow open ditches to carry away excess rainwater beside the maintained trackways. Wade's successor, Major Caulfield, completed a very much more complex network of roads to the same high standard. While this structure was suitable for the marching of soldiers, the passage of carriages or carts with solid wooden wheels, which continued in use until at least the 1790s, quickly disturbed and rutted the surface. The widely used 'Scotch Cart' drawn by two horses would commonly carry a load of one and a half tons if the road was in good condition, normally the case in summer. Despite the problems of surface rutting, the same method of road construction was followed by road makers elsewhere, notably by John Louden McAdam.

The early tracks in Fife followed ancient paths, many of which were not suitable for horse-drawn carts and fresh routes were developed, taking lesser gradients around, rather than directly over, many of the hills. The Moor Road over the Riggins between Cupar and Ceres is readily traversed on foot or on horseback today, but is not suitable for heavier traffic. However, a later, but longer, alternative road today provides gentler but still varying gradients better suited to draught horses.

In the 1730s the road from Crail to the Tay ferries was upgraded and served for coal transport as it passed the Greigston workings. Likewise, a route from Ceres passed the Callange coalfield towards the Ladeddie and Drumcarro coal mines, which remained in operation until the late 1860s.

Although the first parliamentary act permitting construction of turnpike roads was passed in 1663, the first turnpike in Fife, from Queensferry via Kinross to Perth, opened in 1753, a delay of a mere 90 years. The second Turnpike Act in 1790 listed a possible 17 roads that were deemed suitable for upgrading in Fife, only five of which were believed to have been completed by 1800 (Silver, 1987).

The earliest known tollbar in Fife was opened at Struthers in 1801. However, the toll at the Kirkforthar Junction (New Inn), midway between Cupar and Gallatown, had the highest income of all such structures in Fife following earlier road improvements to the south by the collier lords St Clair and Rothes.

The proliferation of secondary roads during the nineteenth century reflected the growing interest in the mining industry, with many minor roads serving isolated workings across the east of Fife. Most such routes were short-lived, remaining active for no more than a few years, as did the collieries.

From the early 1700s, in response to the very poor conditions for transporting coals along the roads of the day, several mine owners developed wooden wagon ways along which horses could haul tubs of coals to nearby ports. The systems at Fordell and Pittenweem were particularly well developed, but they were by no means alone. Later the horses were replaced by steam power, and iron rails were substituted for the wooden ones, speeding the delivery of the minerals to the ports.

The Railways of Fife

In 1845 steam railway systems were rapidly expanding throughout the country. Various proposals were put forward to provide links between the principal towns and the main Fife-side ports on the Forth at Anstruther, Kirkcaldy, Burntisland and Alloa.

Of 15 such schemes submitted, only that proposed by the Edinburgh and Northern Railway, established and financed by the local agricultural estates, was approved in recognition of the potential improvement of transport of the agricultural and mineral produce from their estates by railways. The project was funded locally in the face of disinterest from both Edinburgh and Dundee. The pioneer passenger service in Fife opened in 1847 between Burntisland and Cupar.

From its terminal at Burntisland, the line led to Cupar by way of Ladybank with a proposed branch line development to Perth by way of Newburgh. The proposal to use a bridge at Mugdrum Island to cross the Tay was vetoed by the Admiralty and Perth Harbour Authorities and an alternative route to Perth from Newburgh via Bridge of Earn was selected. Little more than 28% of the revenues were expected to be generated from the transport of agricultural and mineral products.

The railway from Burntisland to Cupar was opened in September 1847. In June 1847 the Edinburgh and Northern Railway bought the harbour at Burntisland and the seven ferries plying to Granton. In May 1848 a line between Cupar and Ferryport-on-Craig (Tayport) was completed and the Company purchased the *Express*, a 269 ton paddle steamer which provided a service to Broughty Ferry until

1879. In 1849 the mileage rates for mineral traffic by rail were set at 6d per ton, with special class goods at 7½d per ton and all other classes of materials at 1/- per ton.

Meanwhile Thomas Bouch, the Company's manager and engineer, initiated the concept of a 'floating railway' across the Forth whereby wagons containing freight, including coal, could be loaded directly on to two new ferries by way of inclined rail-bearing slipways. The seaward end could be raised or lowered as required by the tidal conditions to permit the wagons to be pushed on to one of three tracks on the deck of the vessel. The double-ended 389 ton *Leviathan,* which entered service across the Forth in 1850, could carry up to 34 wagons on each journey. A second one-ended train ferry, the *Robert Napier,* capable of carrying 18 wagons on two tracks, entered service on the Tay in 1851, sailing from a new harbour at Tayport (the renamed Ferryport-on-Craig). The initial build-up of revenues from the business expanded satisfactorily, the minerals, merchandise and parcels traffic revenues rising from £62,250 in 1852 to £103,069 in 1856.

In addition to ever-increasing demand for coal supplies from heavy industry during the Industrial Revolution of the late eighteenth and early nineteenth centuries, there was also a great expansion in agricultural activities, stimulating a substantial demand for the limestone products used in the improvement of soil quality. Many small limestone quarries and large kilns (replacing the earlier sous kills), had been developed across Fife, most producing small quantities of material principally for local use. Some of the workings were more substantial, also taking coals from associated seams for heating the kilns used to render the materials suitable for agricultural use. At Cults near Cupar, a private railway transported the produce to the main line station at Springfield. In the early 1840s the average annual production of limestone was over 6700 tons, more than 2000 tons of which were carried by carts to Newburgh as the nearest exporting port.

For many of the carters, the income from carrying the coals during the winter months enabled their small farms to be kept going during the rest of the year. The cost of overland cartage served to keep the price of the coals to the general public high. In the 1836 New Statistical Account of Abdie, the minister bemoaned the fact that irrespective of whether the coals on sale were from Balbirnie or maritime imports, the price was identical at 10/- per cart load. He correctly foresaw the availability of cheaper coals for his parishioners if railway transport could be provided. The Edinburgh Coal Company calculated that coal would cost as much as 7/- per ton less at Perth and 3/- per ton less at the Forth ports once a railway connection was operating. In actuality, not only were the coals cheaper, but the prices of agricultural produce as a whole similarly fell as the railways constructed suitable sidings for the handling of a wide range of goods once the stations had been established. Between 1846 and 1860 following the development of the railways, the toll collections on the Fife turnpike roads fell by 40% reflecting the importance of the change from road to rail (Bruce, 1980).

By the mid-1800s coal production had increased dramatically, especially in western and central Fife, and by 1855 it was reported that in the previous half-year alone, coal transport on the railway had increased by about 33,000 tons above that of the preceding equivalent period. The Edinburgh, Perth and Dundee Railway Company had rebuilt about one-third of its stock of 1500 wagons. New wagons with greater capacity than their earlier counterparts had been bought. By 1861 the supply of wagons was not sufficient during periods of high demand. They were, of course, still being taken by ferry across the Forth. In the final half-yearly report of January 1862, it was recorded that 369,000 tons of coal had been transported over the 78 miles (125.5km) of rail track and 42,460 wagons had been ferried across the Forth.

Delays in shipping the coal were frequent, and a major extension to the Burntisland harbour in 1860 was timeous as the scale and importance of the Fife coal industry were becoming recognized as the decade advanced. In September 1872 some 62 vessels from Germany or Denmark lay within the harbour awaiting coal cargoes; some had already waited for more than six weeks. Further harbour extensions were built and hydraulic loading cranes installed, allowing 1000 tons of coal to be loaded on to the ships per hour. In 1877 railway wagons carried 287,000 tons of coal from Kirkcaldy to Burntisland, but by 1885 this total had risen to over 770,000 tons. The shipment of coals did not drop away immediately following the opening of the Forth Railway Bridge in 1890, but the demise of the ferries across the Forth yielded additional space for the international shipping seeking the exported coals. By 1905 the cargo totals reached 1,817,000 tons, rising to 2,431,000 tons in 1911. A very significant part of these loads was destined for the German navy.

In 1887 the first dock at Methil, complete with three hydraulic cranes, was opened principally to take part in shipping the produce from the expanding Fife coalfields, and this eastern port rapidly overtook Burntisland in importance. A second outer dock, completed in 1900, helped to raise the annual coal traffic to nearly 3,000,000 tons by 1910. Further permission was given to the Wemyss estate to develop the Wemyss Private Railway to service its collieries, which alone sent over 2,000,000 tons of coal to the dock by 1913. In 1923 the vessels collected a total of 3,368,000 tons of coal from Methil.

Round the East Neuk

The first segment of the Fife Coastal route between Thornton and Leven opened in 1854. Shortly thereafter Sir Ralph A. Anstruther of Balcaskie called a meeting to address the possibility of an extension from Leven to Anstruther, which Thomas Bouch had estimated would cost £58,000. John Wood, banker from Colinsburgh, concluded that in addition to the transport of grain there would be considerable agricultural demand for the transport of potatoes and cattle. He also referred to the importance of carrying coals, which were often difficult to obtain. (That seems to have implied import rather than export of the coals). Furthermore, there was a very substantial trade in cured herring being taken to England. This all suggested that the extension would be economically viable.

The Leven and East of Fife Railway opened a goods line to Largo and extended it to Kilconquhar in 1857, whence to Anstruther Wester in 1863. In 1883 the line was carried forward to Crail and Boarhills before its completion to the new station at St Andrews in 1887. In 1885 an average of 30 tons of agricultural produce and general merchandise travelled west each day along the single track railway which had passing loops at stations and special sidings at the towns en route. At Anstruther there was room to park up to 40 goods wagons to be used during the herring fishing season. In February 1886 during the peak of the season, 190 wagons left the East Neuk on one day, 174 on the following day, and more on the third day. This was a very profitable line during the herring season before over-fishing led to the industry failing.

The East Fife Central Railway extended from Leven through stations at Kennoway, Montrave and Largoward to its terminus at Lochty. Traffic was confined to goods working and limited to a maximum speed of 25 miles per hour. It was the longest goods-only railway line in Scotland, as the North British Railway considered that the population of the area served by the line was too small to justify a passenger service. The goods sidings constructed at all four stations continued operating until 1964. The eastern extension of this line was encouraged by the results of drilling for coals in the Lochty area but the seams proved discontinuous and the mining was short-lived.

Bruce (1980) quotes Mr J.M. Bennett, former station master at Largoward, as saying that seed potato traffic and grain were the main sources of revenue. For some years, coal from Radernie Colliery and from other nearby collieries was transported to Methil docks. The colliery transport driver was said to prefer to work with the 6 ton wagons because they made for easy tipping of his loads.

The St Andrews Railway was a 4.5 mile (7.2km) link connecting St Andrews to Leuchars, opened in 1852, which more than achieved the impact foreseen in the prospectus for its formation. This had projected that, while 66.8% of the traffic would be derived from passenger transport, the minerals industries would account for over 21% of the traffic in the form of coals, lime, pavement and hard stones, freestone and tiles. The Fife Herald of 21 August 1851 reported the suggestion that the predicted transport of coals to St Andrews from coalfields in the west [of Scotland] and also the freestone transport from the Strathkinness, Nydie and Kincaple areas had been underestimated. In due course this suggestion proved to have been fully justified.

According to Bruce (1980), at the time of their greatest development, there were 234 miles (378km) of track and 75 stations within the Kingdom. However, following implementation of the Beeching recommendations, only 81 passenger miles (130km) of railway track remained with 18 stations. Both passenger and goods traffic have since fallen, partly as a result of the opening of the Tay and Forth Road Bridges, partly due to the closure of many of the coal mines after Nationalization and the later demise of the coal-fired power stations. Today only the main line from the Forth Railway Bridge and Burntisland to Cupar remains, with its extensions to Perth via Newburgh and to Dundee across the Tay Railway Bridge.

Chapter 8

The main methods of coal extraction

The techniques employed for the removal of the coals began in simple form and increased in complexity and efficiency with time. This is not the place for a detailed appraisal of the evolution of the methods used, but a brief outline will provide a background appreciation of the industry.

In the earliest days of seeking the coals, the coastal cliffs provided the simplest means of accessing any available coal seams that could be seen readily from the cliff-foot beaches and the coals were removed as tidal conditions allowed. Two examples of this form of deposit are to be found near Crail, the first around Roome Bay to the east of the town and the other to the west of the harbour. In the sixteenth century, the former gently inclined seam is said to have been traced around the head of the bay to return to the coast east of the harbour. To the west of the harbour, beyond the headland, the cliffs host a thin seam that was also later worked a short way inland at Kirkmay Farm. Some old cliff-foot workings are present at Witch Lake, St Andrews but it was not physically possible to exploit the Harbour Coal in the central fairway into St Monans Harbour.

One of the most important factors needed for the understanding of inland coal seams and their structure, which helps to define whether the coals may be profitably worked, is an appreciation of the fact that the seams are simple three-dimensional planes. Their distributions may be readily understood from the fact that the planes may be tilted, folded or torn. A coal seam is very similar to a sheet of paper, extending over a broad area while the layer itself is very thin. Like sheets of paper, given the application of the appropriate forces, the layers of a succession of rocks that includes the coals may take up any orientation from horizontal to gently or steeply inclined, vertical or even overturned. They may become gently folded to form shallow basins and arches or deformed into complex, even inverted patterns. In addition, they may become torn apart along faults that are natural fracture planes. The possible geometric patterns which they may assume are manifold and the patterns of outcrops intersecting an irregular land surface add further complications. The biggest challenge to the would-be miner seeking to exploit coal in a newly found site is to gain an early understanding of the positions of coal seams within the sequence of the rocks and then to discover the ways in which they may have been disturbed by geological processes. Only then is it appropriate to select a method to allow the resource to be accurately located below ground, and plans drawn up to formulate a strategy for working the deposit.

River-bank exposures of outcropping coals are normally relatively restricted in size but give indications of where the coals might be found outside the valley floor and slopes. Few such sites are available to be seen in the East Neuk today. The nearest locality where this form of exposure is to be seen is at Letham Glen along the Scoonie Burn on the eastern outskirts of Leven. Here beside the Scoonie and Sillerhole roundabout on the A915 road, a well-built track leads past several steeply inclined coal seams that have been worked in the past. Their locations are indicated on information boards near the entrance to the glen.

Unfortunately, in the east of Fife much of the ground surface away from the streams is covered with deposits of sand and gravel left behind by the waters escaping from the melting ice sheets towards the end of the recent glacial period. This makes locating the positions of many coal seams difficult to achieve. During many stages of exploration for inland coal seams it was necessary to dig through the superficial deposits, which are commonly more than 3m deep, to seek the seams hidden beneath them. In the sixteenth and seventeenth centuries this form of exploration provided work undertaken by the formally termed 'ground breakers' (the original use of the over-worked term of today). Where successful, the ground breakers pinpointed locations at which the coals were present and followed the patterns of the outcrops of the coals beneath the cover of sands and gravels, peats or soils across the land surface. The outcrop is not the same as an exposure of the coal. At an exposure, the coal is readily available to be observed and measured in place without recourse to the need to remove the overburden. In practice, the boundaries marked on maps by the two methods may differ on the surface by several metres.

If the seam is inclined regularly, the spacing and heights of no more than three of the defined outcrop or exposure points are normally sufficient to allow calculation of the dip of the seam from the horizontal and the angle towards which it is inclined. Thereafter, planning of the system for its possible exploitation may be undertaken.

The early excavation of coal from any point on the land surface requires the removal of material from above the outcrop of the coal seam, creating a hollow or depression on the land surface. By extending the hollow along the face of the newly exposed seam and cutting into it, the coal is accessed and released to be taken away. The expansion of the hollow along the outcrop creates a ditch in front of a new working face. A low ridge on the outer side of the ditch is formed initially by the displaced residue of the glacial deposits and later by waste materials displaced from the tunnels excavated into the coals. The openings or portals into the tunnels were normally left open, creating the appearance of eye sockets peering from the rocks. As a result, they were often referred to as 'ingaun e'es'. In the early days of working such sites, the ditch was commonly left open to the elements as the digging progressed into the hill, as described by Muirhead (1792), the Minister from Dysart, in the First Statistical Account. According to Mr J.M. Bennett (quoted by Bruce, 1980) this three-mile-long ditch remained scarring the land surface near Dysart for at least a century after its creation.

Cutting a tunnel or roadway into an almost horizontal or gently inclined seam to release the coals creates an 'adit', originally not much wider than the shoulders of the hewer himself as he ensures that the tunnel roof remains stable. The splitting and breaking of the coals releases any natural gases from within the coal. If the concentration of these gases builds up sufficiently, the working may become hazardous to health and the hewer could be overcome by 'the stifle'. He would normally recover if taken into the open air once more. If the gases released are small in quantity and not toxic, the working will progress until it reaches the water table, an underground level beyond which the rocks are saturated with water and are increasingly difficult to work.

If a second entrance or adit is cut into the outcrop parallel to, but some metres away from, the first adit, the quantities of coal produced will be similar. However, if the two tunnels are interlinked along the seam itself, additional coals will be released from the rocks. Expansion of the width of the linking tunnels will give rise to the development of 'rooms' if the strength of the roof material above the seam is sufficient. Removal of coals from the rooms yields substantially more coal than from the tunnels alone. The size of the rooms created is strictly limited according to the characteristics of the roof rocks and the availability of timber props to support the roof as excavation progresses. Where the overburden is relatively small, the rooms are normally several metres in width and length, but when the seams are followed to areas with deeper burial, the dimensions of the rooms are maintained, but the widths of the walls or pillars ('stoops') separating them are increased on the grounds of safety. In most situations, more coal was left in the stoops than was extracted from the rooms, so this is not the most economical means of working a seam. Where the early room and stoop working was at shallow depth, modern excavation near Halbeath during the development of the M90 motorway across the west of Fife was sufficiently deep to reveal the spacings of the pillars and the limited areas of the intervening rooms in such methods of working (Figure 8.1). In fortunate areas where seam continuity is not greatly disturbed by folding or faulting, the areas excavated may become very substantial and the seams worked are sometimes quoted as having given access to so many acres or hectares. Few deposits of this undisturbed nature occur in the East Neuk, where the challenges to the workings are much more complex.

Figure 8.1 The tops of a near-surface Room and Stoop mine working revealed during construction of Fife Regional Road c.1980. (Photograph: J. Parry, Deputy Roads Engineer, Fife Council.)

A mine in which the working tunnel starts at the ground surface and follows the seam downwards, as at Radernie, is referred to as a 'dook' or 'drift' mine. As the tunnels penetrate deeper and deeper into the ground the rocks become progressively soaked with naturally accumulating groundwater. The upper level at which the rocks become saturated is referred to as the 'water table', which rises and falls in response to the rainfall. It is not possible to work the coals in thoroughly saturated material. The deeper the penetration of the dooks into the coals, the greater the problems from the groundwaters faced by the operators. The removal of the water was initially achieved by the lowering of buckets down shafts to the seams and raising them to the surface by 'winless men' who powered hand-driven windlasses. However, this system could remove only limited volumes of water and an alternative, less manpower-demanding, longer-term solution to draining the unwanted waters was to construct a separate tunnel within the rocks to lead the waters away from the area to be worked and down slope into a local stream valley. This drainage tunnel, which may be several kilometres in length, is known as the 'day level' and the workable overlying rocks drained by it are said to be 'level free'. The day level leading southwards past Brown's Pit and Blakelyhill on the eastern outskirts of Largoward Village flowed south through Balcarres Den as part of the headwaters of the Kilconqhar Burn. Known as the Long Mine, this linked a series of shallow workings running from north to south through the lands of Largoward (Fig. 8.2 after Mitchell, 1778). A second day level in the Cassingray area is marked on map compilations by Forbes (1955) as starting beside Ure's

Figure 8.2 Mitchell's Map of Largoward Coalfield, 1778. The dotted line indicates day levels.

Pits about 160m north of the former railway embankment and tracking westward for 202m until it passed beneath the railway line, the mouth of the level being clearly marked where the waters were discharged on the south side of the embankment.

Away from the streams and valley sides where the coal seams were at relatively shallow depths, it was possible to reach the coals by cutting vertical shafts from the surface. These shafts, often as little as 2.5m in diameter, led 15m to 25m downwards through the covering sedimentary rocks, which had greatly varying strengths, before reaching the coals. In the Ceres coalfield, the Banfield Coal Book shows that in the period 1749–1761 fresh shafts 40m to 50m deep were cut to the coals of the Lower Limestone Formation every six months or so. While the access to the level-free coals was well defined by the shafts, the distance from the foot of the shaft from which the coals could be worked was controlled by the stability of the overlying sedimentary rocks forming the shaft margins and the roof to the workings. The coals were taken from progressively increasing radii from the shaft foot until excessive shaft and roof collapse rendered the workings dangerous. This form of mine was known as a 'bell pit', as it widened rapidly towards the base of the shaft (Fig. 8.3). There can be little doubt that many of the workings in the East Neuk began in

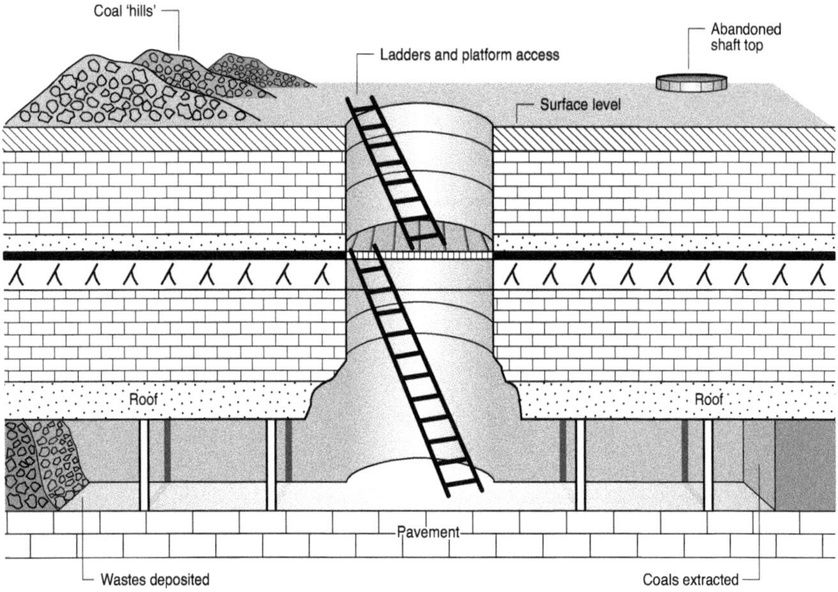

Figure 8.3 Vertical section through a Bell Pit worked with ladders. (Figure prepared by Graeme Sandeman.)

this form. In shallow pits the access and egress to and from the workings was achieved by ladders, sometimes single, but by multiple ladders with intervening platforms where greater depths were required.

One of the best examples of bell-pit shaft arrangements and spacings to be recognized today is in the Brewsterwells March coalfield at Higham Toll beside the main St Andrews to Largo road (A915). Little can actually be seen on the field surface from the ground, but a satellite image from Google Earth, taken while the fields were being harrowed, provides an exceptionally clear picture of coal debris on the north-westward dipping southern limb of the Radernie Syncline to the southeast of the main road junction (Figure 8.4). Scaling of imagery obtained by aircraft or satellite is often difficult to determine, but in addition to the buildings and roads within the area of the coalfield, a red tractor can be identified towing the harrow. The dark circular spots near the centre of the image mark the residues of coals deposited on the field surface around the shaft tops by the bearers who had carried to the surface the coals won by the hewers below ground. The circles are aligned in three 150m long strips (about half the width of the field), trending northeast to southwest across the field with three or four circles forming each line. The cluster of circles is thought to mark the positions of the three uppermost coals. To the southeast there follows a gap before a fainter cluster of circles marks the position of the few workings on the Radernie Brassie Coal 11m or so below the Radernie

Figure 8.4 Brewsterwells field at Higham crossroads showing tops of many bell pits and mine shafts (Google Earth image).

Duffo Coal. A dark line 3m wide crosses the field towards the main road in the northwest, where it joins a fan-shaped feature formed beside the present field exit and presumably marks the position where the formerly transported coals were carried to the main road. Between the road and the north-western line of dark circles, three or four faint circles along a distinct parallel linear feature mark the line of the Charlestown Main Limestone. There is no sign of the location of the St Monance Brecciated Limestone, presumably to the southeast of the woodland area (After Mitchell, 1778).

None of the lines marks the actual position of the outcrop of the coals or limestone at the surface. Each indicates the line at which the Charlestown Main Limestone or one of the four Radernie coals is present at the 20m to 30m depth reached by the shafts. The true outcrops would be displaced 60m to the southeast if the dips are 14° (gradient of 1 in 4) to the northwest, as suggested by Terris (undated). The outcrop of the St Monance Brecciated Limestone (below the coals) is not as readily determined, but according to the BGS geological map, should pass through the dark area to the southeast of the tractor. Plans of the three coal seams worked in this field at different times are shown in Figure 8.5. All three seams encountered coarsening of the coals along a similar, almost west–east, line in the southwest before reaching the line of the Crail road (B940), suggesting the presence of a fault or dyke below ground.

The construction of multiple bell pits to exploit the coals at any locality is expensive in terms of labour and wasteful of resources in that it results in less than half of the available coal being recovered. Where the presence

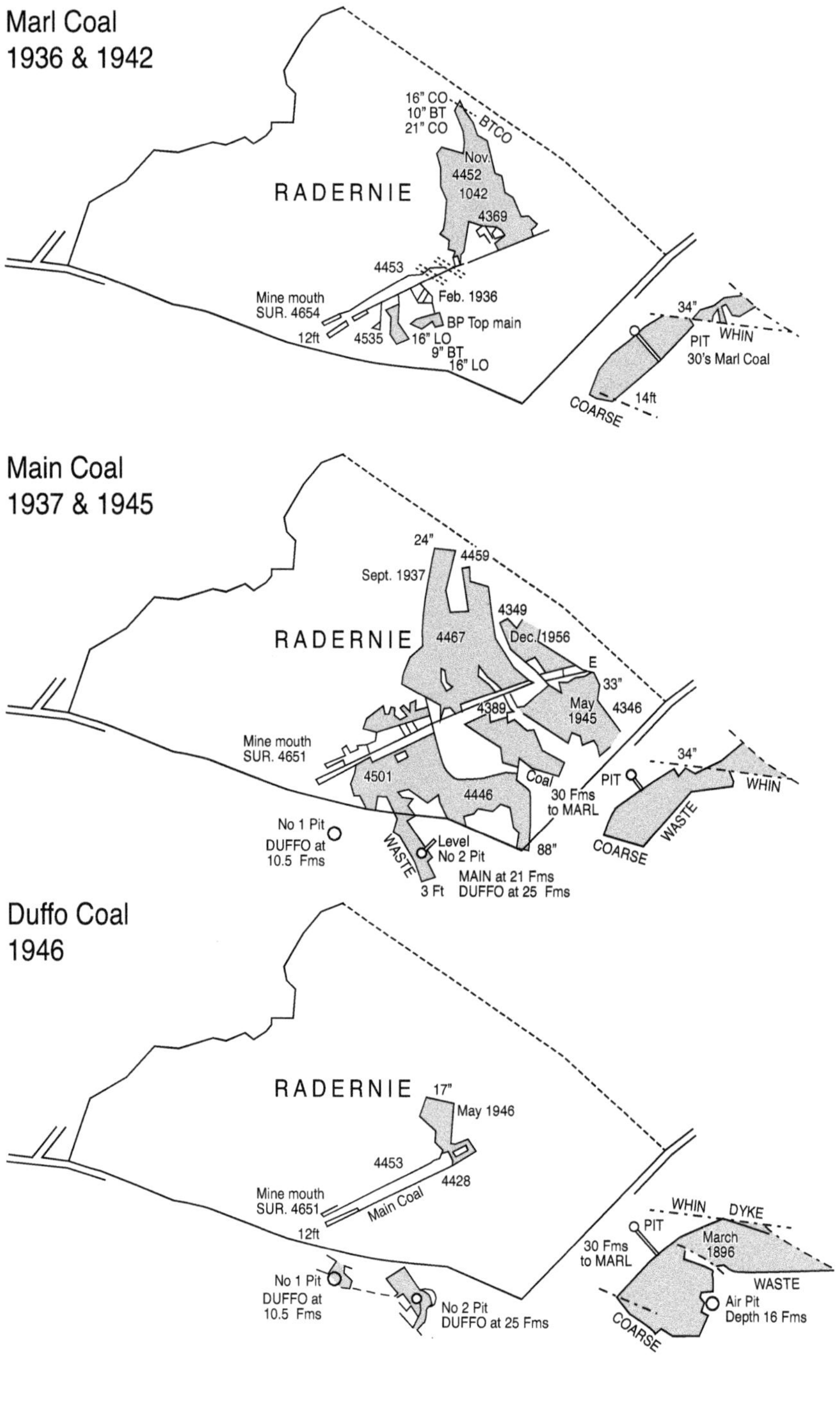

Figure 8.5 Selected plans of extraction progress on the Main, Marl and Duffo coals of Radernie, 1896, 1936–7, and 1942–46. (Source: mineral evaluation survey, adapted by Graeme Sandeman.)

of substantial coals at greater depth had been demonstrated by boreholes, broader diameter circular, or less commonly, square shafts were installed and taken to greater depths than were possible or sensible for bell pits. Such shafts, penetrating to greater depths, required improved roof stability for the personnel and securely fixed wooden stairways with intervening platforms for the drawers and bearers. Many of the shafts were lined with brick or timber for safety. Access to the coals by ladders or stairways was soon to be followed by more direct access by means of winding ropes to raise and lower tubs of coal, unwanted waters and even personnel. All solid waste material was retained underground.

The problem of how to rid the workings of the groundwaters became increasingly pressing with the greater depth of working. The manually operated windlass-powered extraction systems (Figure 8.6A) were replaced by those driven by water wheels powered by local stream diversions (lades) as constructed by the Earl of Kellie on the Dreel Burn north of Waterless Bridge in the Pittenweem coalfields (Figure 8.6B). Wind-powered pumps were less reliable, being heavily dependent

Figure 8.6A Manual operation of a windlass pump to unwater the mine (Google image: www.gb.fotolibra.com/images).

Figure 8.6B Stream-powered overshot water wheel method of unwatering the mine (Google image: www.fotolibra.com/gallery).

upon local weather conditions. More reliably, horses provided an improvement to the drainage problems by powering 'gin pits' (engine pits) using single horses or teams of horses to drive winding wheels to raise waters on chains or buckets to the surface (Figure 8.6C). Later came the introduction of the Newcomen steam-powered 'atmospheric' pumps in the early eighteenth century and later still, the more economical and advanced Watt steam engines after 1760.

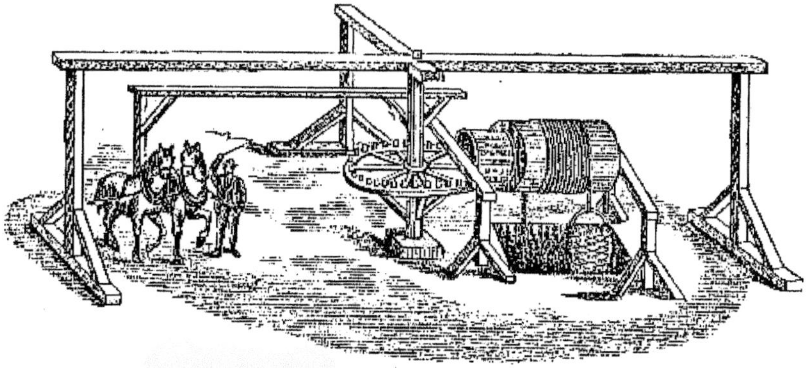

Figure 8.6C Horse gin: another method of unwatering the mine (Google image: www.scottishmining.co.uk).

At the foot of the shaft, coals were largely left untouched to act as foundations and to ensure the stability of the shaft and shaft foot area, where the initial coal collection and processing activities would take place. From this central point a major horizontal roadway or gallery was established along and within the main coal seam, with the seam continuing to rise above this level on one side and to fall away on the other. At most sites the coals would be worked firstly in rooms extending up the rising side, with the intervening walls or pillars being increased in thickness with depth below the surface.

Stoop and room, sometimes called pillar and room, workings were those most commonly worked in eastern Fife. The coal extraction was organized in a regular fashion so that the stoops or pillars were lined up, leaving straight passages for access with gaps cutting through the lines (see also Figure 8.1). Where appropriate, dooks, sloping tunnels leading down dip from the main gallery, were cut to create subsidiary roads below the level of the shaft foot. From all of the open areas, the coals were cut by the hewers (Figure 8.7) and removed down slope to the road or gallery for transport to the shaft foot. This was the task for the putters and bearers (Figure 8.8A) who moved the coals in wooden tubs along the roadway floor. Later these tubs or boxes were pushed or pulled along wooden rails like sledges. Eventually the tubs were given wheels before iron rails were laid in the main galleries (Figure 8.8B). Later still, if the gallery roof was sufficiently high for the putters to walk upright, pit ponies were introduced to pull the tubs in some workings. It is important here to realize that few of the coal seams in the East Neuk were as much as 1m in thickness, and the roof of the road normally required to be lifted and supported by timber cross-members to enable the colliers' assistants to walk with their loaded tubs of coal (Figure 8.9).

In workings where several coals are arranged above each other, it was normal for the uppermost pillars left standing to be smaller than those at

Figure 8.7 The Hewer: miner using a double-pointed pick for handworking thin coals (painting by Derek Slater).

Figure 8.8A Woman bearer dragging 2cwt coal tub to the foot of the shaft to be lifted to the surface *c.*1720 (1842 Government Inquiry: Employment of Women in Mines).

greater depths below the surface. The chessboard format of the pillars and intervening rooms provided a means whereby the flow of air through the workings could be regulated by means of simple hinged doors that could be opened or closed as required. This method of operating permitted only a fraction of the

Figure 8.8B Woman and two children dragging wheeled coal tub to the foot of the shaft c.1780 (Google image: www.industrialrevolution.org.uk).

Figure 8.9 Filling the tubs: modern tubs conveying the coals to the shaft for lifting to the surface. (*The Loading Point*, painting by Derek Slater.)

available coals to be removed from the seam, and in many cases only half of the coal was removed. As the workings were approaching abandonment, the pillars were commonly at least partly removed.

With increasing distance from the main shaft, the quality of the air decreased through breathing by the colliers and also the release of natural gases from the coals. This, coupled with coal dust in the air, led to severe health problems for the colliers, and in most mines at least one separate air shaft (often the escape or bye pit) was created to allow fresh air into the workings. Following the introduction of boilers to provide steam power within the mine, the fires and furnaces were placed beside the base of the main shaft. The air and gases within the pit were guided along the roads and galleries towards the shaft foot, where the warmed escaping dusts and gases were encouraged to pass up the shaft, causing stale air to be expelled and fresh air to be drawn into the mine from the remote air shaft.

There was work for support teams checking the tunnel walls and roof for stability to maintain the through roadways for staff and materials. Timber props were inserted in many of the roofs (Figure 8.10) and these were later supplanted by steel girders where appropriate (Figure 8.11). Until adequate ventilation became available, remanent pockets of gas provided danger and explosions triggered by conditions. Bings, heaps of waste material, sometimes exceeding 50m in height, began to build near the pit head. Some of these black spots on the landscape were indeed built in the East Neuk after 1842. As many of the workings were approaching closure, there was little opportunity for bings to reach great size. After the collieries closed, the bings were removed, one of the last disappearing from Largoward in 1973. Some bing remnants still occur at Radernie. A great expansion of coal production had taken place locally in response to industrial demand during the nineteenth century, leading to the

Figure 8.10 Fixing the beam: colliers erecting timber roofing supports (painting by Derek Slater).

Figure 8.11 Assembling the girders: provision of steel girders to support gallery and roadway roof (painting by Derek Slater).

inevitable exhaustion of the reserves and mine closures. The flames from tallow lamps or candle-powered lighting were distressingly common features of many coalfields, with many resultant deaths through explosions. The development of the safety lamp by chemist Sir Humphry Davy in 1815, stimulated by the demands of miners of the Northumberland–Durham coalfield, marked a great step forward in mine safety. Davy did not manufacture the lamps himself, but provided his design for several companies throughout the British coalfields. Many of the colliers had their own personal safety lamps, but there were still explosions and over 1000 people died underground annually in the 1860s, rising to 2000 a year by the turn of the century (Montgomery, 1994). The recognition that the coal dusts were also major contributors to the problems led to further improvements in the evolution of mine safety (Figure 8.12).

A further development of safety equipment came with the creation of the carbide lamp, which was produced in several forms. The principle on which it worked was the release of drops of water onto pellets of calcium carbide to give rise to acetylene gas, which burned with a soft but powerful light. The best-known form was worn on the miner's hard helmet and was capable of a secure beam of light able to last a full shift. A second form of lamp could be placed on a ledge or hung from supports, and still others were hand-held. Introduced about 1900, the carbide lamp was powered by pellets bought in Fife at the local

Figure 8.12 The Face: two colliers working on the coal face with the area lit by safety lamps (painting by Derek Slater).

co-operative stores on Saturday mornings, often by children of the family to earn their pocket money.

A much more economical system of coal working using bell pits or stoop and room operations is the longwall mining system introduced to the east of Fife under the name of the 'Shropshire Method' whereby the entire seam was removed in a single excavation process. A long 'coal face' or wall was cut along the seam and the entire seam of coal was removed systematically without leaving a permanent pillar (Fig. 8.13A, Figure 8.13B). This is the longwall working method.

The coals were worked into the exposed face, effectively removing slice after slice of the coal along it. Behind the colliers, the roof was supported by props of timber, or later by hydraulic units. As the face was pushed back, the early props were removed and brought forward into the new working area, deliberately leaving the gap from which the coals had been removed. In time, the roof of the no longer supported area would 'sit' (collapse) into the gap created as the excavation work continued a short distance away. The longwall mining developed with the coals dislodged from the face originally hand-filling box tubs, but later mechanical cutters allowed the coals to fall directly onto conveyer belts that removed them to tubs and then to the shaft-foot lift. This form of mining could take place with the coal face advancing away from the foot of the shaft or retreating towards it if the workings were started at the far side of the mine beside the air pit. In either case the roof collapse above the extracted

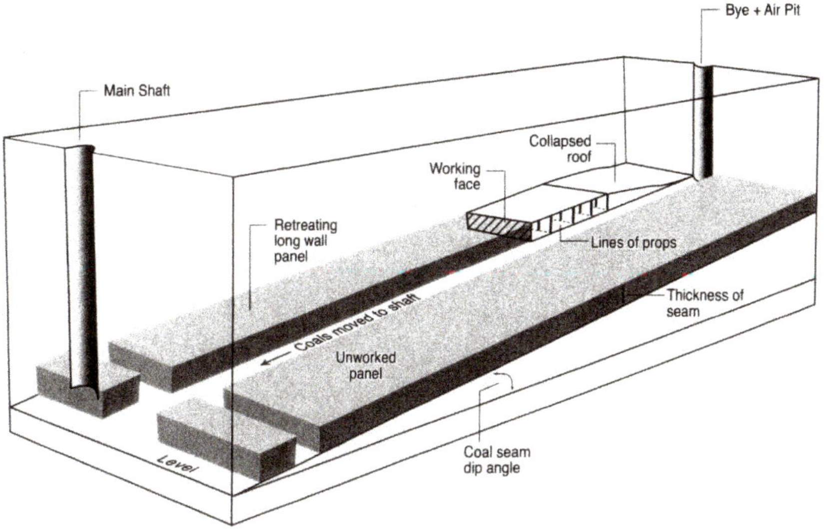

Figure 8.13A Diagram of the working of a longwall mine. (Figure prepared by Graeme Sandeman.)

Figure 8.13B The scraper conveyor: colliers working in a longwall mine and loading the conveyor (painting by Derek Slater).

coal settled relatively quickly, and any expected damage to the land surface above or overlying seams would be calculated in advance to define the breadth of the face to be worked and its projected depth below the surface. With care, no impact at the surface is detected.

Most of the miners preferred to work with wooden props, as they could be heard progressively breaking and splintering before finally giving way, providing the hewer team with time to leave the area. By contrast, failure of the hydraulically operated steel props is sudden and unannounced.

In the early days of mining, the actual cutting of the coals was achieved entirely by hand, with the hewer using one of several tools, commonly either two points or one point and one chisel-shaped end on a single pick haft. There is a natural 'cleat', a regular direction of preferred breaking within the coal seams, which aided the physical effort of breaking the coals free from a working face. Nevertheless, wedges and a range of other devices were used at times to open up the coal and release the most sought-after great coals. From early days of working, the coals were normally holed near the level of the pavement and later small explosive shots were inserted into the holes (Figure 8.14). The charges were fired at the end of a working shift, allowing the dust to settle before the next day of working began. The coals underwent a primary manual sorting by size before the largest blocks, the *great coals,* were loaded into the tubs. Later the smaller *round coals* or *cubical coals* were collected and sent to the surface. Very small pieces of coal suitable for use in lime burning, or later in the furnaces generating steam power, were sorted using a regular 'graip' in which the shovel blade was replaced with a series of bars with slots of defined

Figure 8.14 Preparing the shot: colliers drilling holes at the base of the face for insertion of explosive prior to blasting (painting by Derek Slater).

width. The latter, less valuable, materials were collected separately. As technological advances were introduced, the cutting was achieved with the use of compressed air equipment, and after 1842, the transport of the coals below ground by women and children was eventually replaced by the introduction of powered conveyor belts leading to the foot of the shaft, at which stage the grading by size and the removal of unwanted wastes took place at the sorting tables above ground.

'I remember the heaps of waste materials which covered the land surface of much of West Fife when I was a lad, but I don't remember many, if at all, in the East Neuk, nor the sulphurous smells' – a not uncommon statement when discussing the industry. Before the 1842 Government Enquiry into underground working, which led to laws banning women and children under the age of 12 years from working below ground, the preliminary sorting of the coals according to their quality took place below ground, and the bulk of the wastes were retained below the surface, filling any abandoned rooms left in the old workings. In earlier days the women who were tasked with carrying the coals to the surface would not stand for their loads being rejected on account of too much waste material being incorporated. When the whole mass of the dislodged material was mechanically fed to sorting tables on the surface, the wastes were separated by hand under greatly improved working conditions.

It should never be forgotten that coal mining was a very dangerous occupation. Roof collapses and failure of the roadway walls led to many injuries and deaths. Through time the roof of any major roadway tunnel would have been strengthened either with wooden props or, later, using friction steel props to support both the sides and the roof. The colliers engaged in maintaining the access roads and machinery were vital members of the community. The industry in the East Neuk did not suffer as many losses as in some mines elsewhere, where the thicker seams brought added dangers.

Chapter 9

The geology of the coal-bearing successions

The sedimentary rocks in which the Fife coals occur are all materials deposited during the Carboniferous Period, which lasted for approximately 74 million years between 360 and 286 million years ago. As already indicated, the immediately preceding geological history of Fife had been characterized during the Devonian by the accumulation of rocks of volcanic origin followed by a series of mainly red or yellow coloured sandstones deposited by southward draining rivers in near-coastal conditions. Greensmith (1965) and later Forsyth and Chisholm (1977) showed that the earliest Carboniferous sediments in eastern Fife were very similar, with rivers from the northeast supplying sands and muds to near-coastal environments that underwent repeated marine incursions. The resultant sedimentary rocks of the Strathclyde Group (formerly the Calciferous Sandstone Measures), which host the earliest coals, are overlain by the Lower Limestone Formation, the Limestone Coal Formation and the Upper Limestone Formation sediments, all of which contain some worked limestone and coal seams.

The succession of rocks is not completely straightforward. Some time after the deposition of the sediments, the rocks underwent a series of disruptions leading to their becoming folded on a large scale and developing many fractures with substantial faults breaking across the folded rocks. Late in the geological history several phases of intrusion of igneous material occurred. Some of these pre-dated the final faulting but some of the intrusions are younger. Active volcanoes co-existed with deposition of the sediments at several periods and others, probably linked with the intrusions, created the hundred or so small volcanic necks to be found across the east of Fife. Many of the intrusive rocks have been used as sources of building stones for houses and estate walls, and today active quarries also supply aggregates for road metals.

Effectively the bulk of the East Neuk rocks form the eastern side of a large synclinal (downwarped) fold, the axis of which has been gently tilted south towards Lundin Links and the head of Largo Bay. The oldest rocks encountered are to be found along the eastern coast between Kingsbarns and Fife Ness and the youngest are exposed around Lower Largo (Fig. 9.1). In the core and limbs of the fold, there are smaller folds and fault structures that bring coal seams to the surface. The northern limit of the area under consideration

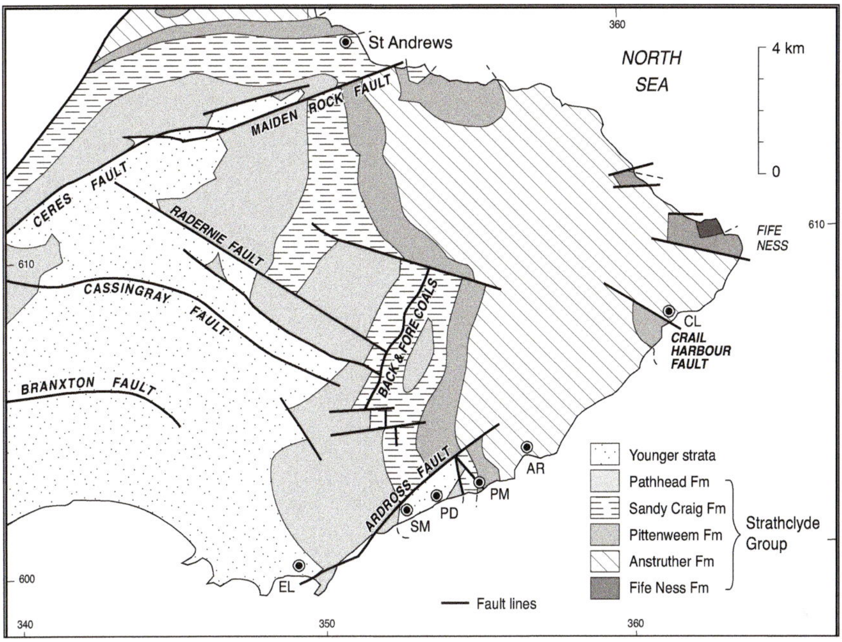

Figure 9.1 Summary map of the distribution of the divisions of the Strathclyde Group (prepared by Graeme Sandeman).

lies along the eastern continuation of the major Ochil Fault, a southwest- to northeast-trending fracture zone with a suspected southward downthrow of between 200m and 300m that maintains its identity as the Ceres Fault leading eastwards to join the Maiden Rock Fault before reaching the coast a short way south of St Andrews. To the south of the Ceres Fault Zone, the open fold through Largoward to Largo is broken by a series of northwest- to southeast-trending normal faults, most of which also downthrow the deposits towards the south in steps each estimated at between 50m and 90m in the coalfield areas (Forsyth and Chisholm, 1977). Extending between the coast at Elie and inland to Kilrenny, the Ardross Fault trends east-north-eastward away from the coast (Francis and Hopgood, 1970). Believed to be parallel to the nearby fracture patterns associated with the Southern Upland Fault beneath the Firth of Forth, this structure predates late Carboniferous volcanic necks that are associated with the fault but are not cut by it.

The Upper Devonian deposits

The Memoirs of the British Geological Survey indicate that rocks of the Upper Devonian period are exposed within the area under discussion. These are at Cambo and Balcomie, near Fife Ness, where they are of interbedded red sandstones and pale-coloured layers of calcium carbonate-rich pebbly conglomerates.

The latter are believed to be of accumulations of materials derived from erosion of soils in which salts were formed during dry periods, as seen today in desert margins. In the lack of fossil contents, the precise age of these deposits remains conjectural.

The Strathclyde Group (formerly the Calciferous Sandstone Measures)

Most of the rocks exposed at the surface in the area from St Andrews to Fife Ness, Anstruther and Elie and east of a line between Elie and Colinsburgh to Strathkinness are from the Strathclyde Group of sediments, as are the sandstones extending south from Craighall Den near Ceres. They were deposited during the rather brief period between 342 and 329 million years ago (Davydov, Korn & Schmitz, 2012).

Towards the top of the Upper Old Red Sandstone the red beds and conglomerates bear the famous *Holoptychius* and *Bothriolepis* fossil fishes of Dura Den (Anderson, 1859). These in turn give way to the increasing layers of sediments of marine origin in which there are signs of cyclically repeated successions of rock types. The desert margins of the Devonian environments of deposition were increasingly subjected to the incursions of the sea. The repeated succession of the marine and non-marine sedimentary rock types are typical of the Strathclyde Group, and host the earliest coal seams of this part of Scotland. Formed of land plants, the coals are commonly associated with thick sandstone units believed to have been deposited by rivers draining across deltas and extending southwards into Fife (Greensmith, 1965).

Exploratory boreholes drilled by the British Geological Survey suggest that the earliest division of Carboniferous rocks, the Strathclyde Group, which form the bulk of the rocks of the East Neuk, may be more than 2000m in thickness. This very substantial succession has been subdivided into five parts: the Fife Ness Formation overlain in turn by the Anstruther Formation, the Pittenweem Formation, the Sandy Craig Formation and the Pathhead Formation (as in Table 1), and as shown on the map provided by Forsyth and Chisholm (1977) and modified in Kassi *et al.* (2004) (Fig. 9.2).

Within these formations two forms of valuable time horizons are recognized. The first of these are the individual coal seams, which may be continuous and can sometimes be followed across broad areas, including adjacent coalfields, or they may be small, isolated remnants of formerly much more extensive seams. The second form of time markers are so-called 'Marine Bands' of muddy sediments and sometimes limestones, which contain readily identifiable and slowly changing fossil organisms such as crinoids, corals, brachiopods and especially the bivalve *Naiadites*. These and the associated thin layers of dolomitic limestones, also of probable marine origin, indicate the start of

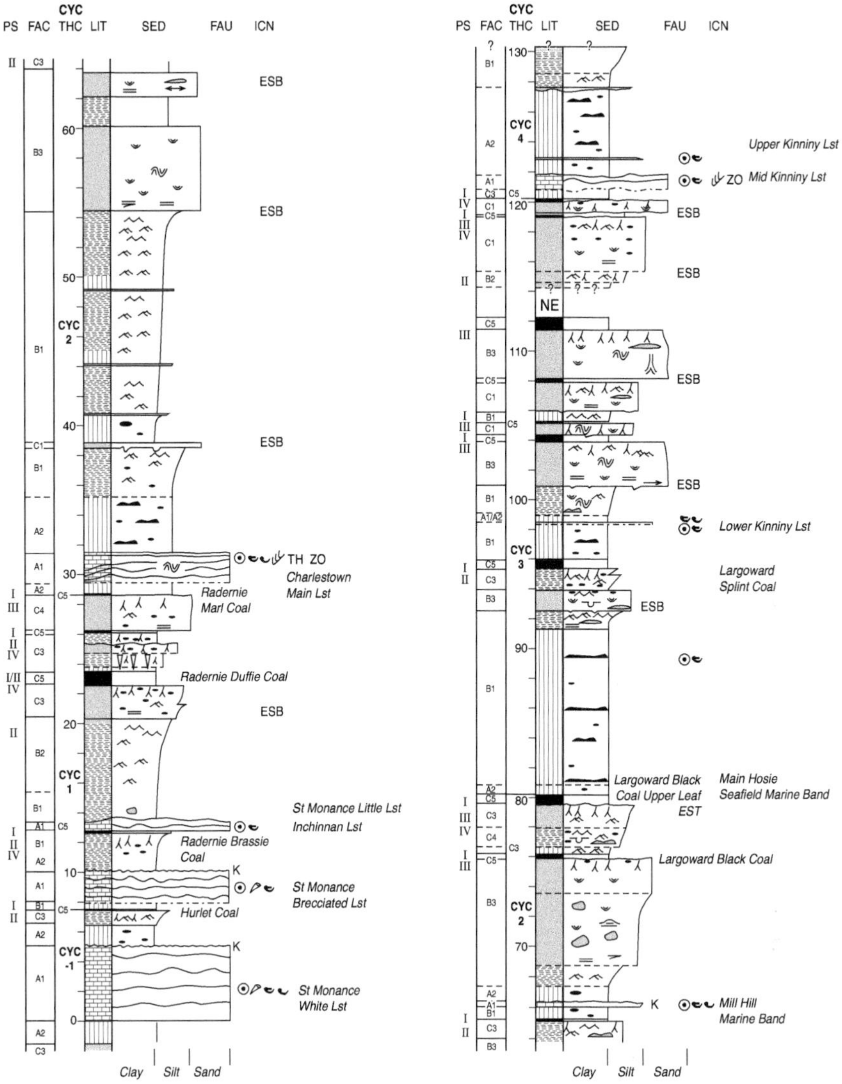

Figure 9.2 Sedimentological section of the Lower Limestone Formation between Pathhead and St Monans (after Kassi *et al.*, 2004: adapted by Graeme Sandeman).

periods of marine conditions in the depositional basin resulting from regional, probably worldwide, rises in sea level. A second series of less widely distributed markers are fine-grained muddy rocks containing non-marine faunas, especially the brachiopod *Lingula* and bivalve *Schizodus* that occur within this succession, each marking a period of extreme flooding by fresh or brackish waters. The changing characteristic morphologies of the fossils in both the marine and non-marine bands may be due to evolution of the populations through time, or possibly due to natural ecological variations of their morphologies in response to geographical changes in water temperature or salinity and nutrient variations along the coasts through time. Whatever the cause, knowing the relative positions of significant coal seams with respect to the other markers provides a valuable guide for coal industry geologists seeking fresh potential sites at which to find and exploit the coals.

The principal rocks of the initial five stratigraphic divisions are generally composed of very similar dark grey or yellow coloured sandstones with darker siltstones and mudstones but in differing relative abundances. Red beds of sandstone, which are common in the Fife Ness Formation, recur in a few layers within the Sandy Craig Formation. Whilst coals are generally scarce in both these units, thin coals are present in the remainder of the succession. Forsyth and Chisholm (1977) suggested the widely recognized *Back* and *Fore Coals* to which Landale (1837) and later Kirkby drew attention (in Geikie, 1902), probably lie within the Pathhead Formation. Gibson (1927) noted that only these two seams were thick enough to work. Formerly worked in the Arncroach–Kellie–Lochty area, the Fore Coal, rarely more than 2m thick, is 40m below the Back Coal, a 2.07m thick seam with three internal shale partings limiting its value. Thin seams of oil shale, similar to those worked in the Lothians, where they are believed to have been deposited in thermally stratified lakes, also outcrop in the Pathhead Formation on the foreshore between Pittenweem and St Monans. Although not known to be extensive in the East Neuk, the oil shales were formerly worked at Pitcorthie, 3km north of Anstruther Easter. At the top of the Pathhead Formation, the Marine Band is supplanted by the St Monance White Limestone, the first of a series of marine limestones which are recognized across the Midland Valley and are used as indicators of the boundaries between defined groups of repeated successions of coal- and limestone-bearing sedimentary rocks.

The Lower Limestone Formation

The Lower Limestone Formation, 329–326.5 million years old (Davydov, Korn & Schmitz, 2012), outcrops from Teasses towards St Andrews, south from Baldinnie to Largoward and the coast at Elie, having particular importance in the Pittenweem to St Monans area. The succession in this part of Fife is cut by many faults, is heavily folded, and also hosts both igneous intrusions and bedded volcanic rocks, which are thin in the north and thicken southwards to 240m in 9km to Radernie.

The base of the Lower Limestone Formation is placed at the bottom of the St Monance Brecciated Limestone (termed the Charlestown Station Limestone in West Fife and the Hurlet Limestone in the west of Scotland). The top of the formation is taken as the base of a red marine mudstone a short way above the *Marl*

Coal in areas where the Upper Kinniny Limestone, which defines the limit in the west of Fife, is absent.

The rocks of the formation occur in a series of consistently repeated successions or cycles from limestones through mudrocks with increasing silt and sand content upwards, and near the top bearing ripples showing the influence of currents and tidal activity during their deposition. Eventually the deposits become dominated by sandstones, essentially former beach deposits or those from the channelled surfaces of deltas. The emergent delta tops and sand banks developed soils able to support plants, giving rise to root-bearing seatearths below what were originally layers of peat, the forerunners of the coal seams. The repeated patterns, which are believed to indicate advances and retreats of the sea, may become interrupted, and sometimes the old delta tops became deeply eroded, such that the sediment deposition cycles are interrupted. The sediment sequences, reflecting what are believed to have been markers of global rises and falls in sea level, are some of the earliest cyclic deposits that characterize the later Carboniferous successions now recognized over vast distances across much of western Europe, Scotland, northern England and also the eastern states of North America. The most readily recognizable boundary between the successive depositional cycles is the Maximum Flooding Surface (Van Wagoner *et al.*, 1988), which marks the end of a period of emergence and erosion. Most commonly this surface occurs at the base of a bed of limestone or a marine band, either of which may rest directly upon a coal seam. (Fielding *et al.*, 1988, Kassi *et al.*, 2004). Individual cycles may vary in thickness from 5m to 60m.

The early parts of the Lower Limestone Formation deposits are thickest in the Radernie–Higham area where sandstones, shales and coals are thicker than at other sites to the east and west, although the limestones are thinner in this central belt. This anomaly suggests that, during deposition, there had been a greater supply of sand and mud to this area, with more rapid basin floor sinking in the central district to provide space for the thick sands, on which four distinctive coals were able to develop. By contrast, in the neighbouring areas where the same coal seams are present, the intervening sediments are much thinner. The coals concerned in upward succession are the *Radernie Brassie, Radernie Duffie, Radernie Main* and *Radernie Marl.* All are of variable thickness, mostly less than 1.2m, and the *Duffie Coal* is locally associated with the Blackband Ironstone, as at Winthank to the north. Two further coals, the *Largoward Black* and 20m above it the *Largoward Splint,* occur higher in the succession between the Charlestown Main and Mid Kinniny Limestones, which are separated by 15m in the south at Muircambus but by 22m only 3km to the north at Lathallan. The top of the succession is marked by the *Marl Coal* a metre or so below the Upper Kinniny Limestone at Lathallan, where it is 70m above the Mid Kinniny Limestone, contrasting with the separation of no more than 38m between the same marker horizons believed to have formed during identical time periods. This difference is sometimes interpreted as indicating that there was differential settlement in sectors of the floor of the depositional basin during the period of accumulation.

The Limestone Coal Formation (329–326.5Ma)

This is the earliest part of the Upper Carboniferous Namurian Succession. Its base is defined as the top of the Upper Kinniny Limestone. The formation outcrops from Shell Bay to Elie before extending inland to Largoward, and thereafter it appears in the New Gilston outlier and in a series of structural fragments along the Ceres and Maiden Rock fault zone, from Ceres to Drumcarro. While the Limestone Coal Formation is dominated by repetitions of cyclic sedimentation patterns, unlike the Lower Limestone Formation it bears few limestone horizons. However, there are several distinct marine bands, mainly of mudrocks (clays and shales) yielding fossils of the brachiopod *Lingula*, an organism that appears to have lived then, as now, in tropical brackish water environments such as estuaries or coastal embayments.

Forsyth and Chisholm (1977) have described the 3–15m thick sedimentary cycles within this formation, starting with coal at the base, passing up into mudstone, siltstone, sandstone and seatearth below the succeeding coal. There are many coal seams, most of which are too thin to have been worked (less than 40cm), but several valuable coals exceed 3m in thickness. In the Ceres area, where coals are said to account for more than 9% of the succession, no fewer than 18 seams have been given names, often reflecting their characteristics, namely: Ceres Thick, Ceres Two Foot, Ceres Six Foot, Mak Him Rich, Ballfield (McManus, 2010). The Limestone Coal Formation also occupies the core of the long-abandoned Pittenweem–St Monans Coalfield in which Geikie (1902 quoting Galloway, 1882) lists 14 coals. The often fragmentary records suggest that there were more coals in the Limestone Coal Formation sediments in the East Neuk than in Central Fife and that these coals provided the basis for the sustained early development of the coal industry in the east of the Kingdom.

There are some volcanic ashes in the lower part of the succession, suggesting that there may have been localized uplifts and depressions of the surface associated with the vulcanicity. However, this has not been fully demonstrated and there is evidence that at least some of the ashes had been injected into the sediments in association with later stronger vulcanicity during deposition of the Upper Limestone Formation.

The principal coalfields of the Limestone Coal Formation were those of the Ceres–Denhead district, the Elie–Largoward district and the core of the St Monans–Pittenweem syncline with lesser small fields such as that at Rires. The former have been examined elsewhere (McManus, 2010). The St Monans workings were subject to disputes between Sir John and Sir Richard Anstruther, who owned different, often adjacent rigs or segments of the land. The long-lived workings here, which had formerly served to provide work for many families during nearly 300 years of operation, initially ceased in 1785, although the final clearance of the structures and workings is said to have extended for nearly 20 years thereafter. It formed the basis of an interesting account by Galloway (1882) to which we will return.

The Upper Limestone Formation

This is not well exposed in the east of Fife. It is believed to occupy an approximately circular outcrop pattern around Largo Law, reaching the coast of Largo Bay at Lower Largo and Shell Bay. Where seen it is mainly of alternations of sandstone, mudstones and seatearths, but it has not been possible to recognize any distinct cyclicity in the deposition. In their discussion of the succession of this formation, Forsyth and Chisholm (1977) found it necessary to depend strongly on records from three boreholes put down by the British Geological Survey. They identified volcanic ashes and tuffs in the lower part of the sequence in two of the logs, but the 1:50,000 scale Geological Survey map shows at least one bed of ash a short way below the Castlecary Limestone at the top of the unit to the east of Upper Largo. The Carhurlie coal is marked as outcropping to the west of the Largo Law volcanic centre.

Chapter 10

The coal seams and mines of the East Neuk

The detail of our current knowledge of the workings in most of the mines is very incomplete and varied. The records are scattered through reports of mining engineers, estate papers, testimonies of former colliers, plans of individual or groups of mines and Mine Abandonment Plans. However, the latter, now held by the Library of the British Geological Survey in Edinburgh, were not required to be submitted until after 1885 and the majority of the workings in this part of Fife had been abandoned well before that date. The archives of the National Coal Authority in Mansfield, Nottinghamshire hold few records from Fife. Only limited records are to be found in estate papers held in the National Archives of Scotland or in the Special Collections section of the Library of the University of St Andrews. Plans of some of the workings are today held by the Institute of Materials, Minerals and Mining at Neville Hall in Newcastle-upon-Tyne. Without exception, the staff in each of these locations have been most helpful in seeking the records and making them available. A map (Frontispiece), including place names of many sites to be examined is provided at the start of this text to enable the locations being discussed to be placed in geographical context by the reader.

Witch Lake and Castle Cliff, St Andrews

In the East Neuk the seam most readily accessed and showing the typical characteristics of coals to advantage is that exposed in the sea cliffs of the Scores at St Andrews (see Figure 10.1). Bordering the Witch Lake to the east of the St Andrews Aquarium and the former Step Rock swimming pool, the cliff provides an excellent introduction. Beware: occasionally small fragments of rock fall from this cliff, which also serves as a nesting site for a colony of fulmars and is legally protected. **Please note that climbing is forbidden and fines are imposed.** However, the features of the coals may be freely observed from the cliff-foot beach under all but extremely high tide conditions. The cliff-foot beach is accessed by way of a sloping path beside the St Andrews Aquarium. Steps formerly giving access to the beach itself are sometimes destroyed during winter storms, and it may be necessary to scramble beneath a short stretch of decking to reach the sands.

At the foot of the cliff, above a 1.5m high concreted wall, is a 25cm thick black coal that is internally layered (Figure 10.1). Below the coals are grey shales that

Figure 10.1 Lowest coal seam in Witch Lake Cliff, St Andrews; bright and dull coals are typical. The former clayey soil holds brown iron-rich nodules. Note sand-filled plant stems at the top of the coals (scale bar 30cm).

formed the soils from which grew the plants giving rise to some of the earliest of the Fife coals. The clays become progressively paler in colour upwards, suggesting that they had become impoverished in the nutrients fed to the plants of the overlying coal from what was formerly a layer of soil now referred to as a seatearth. Step nearer to the rock face and look closely at the lowest part of the dark coal layer. This has a powdery texture that crumbles away under erosion. It passes upwards into a blocky, tough and rather lustrous layer of usable coal less than 10cm in thickness. Near the top of this coal there are a few sections across pale-coloured, circular, flat-lying, sand-filled tubes mostly less than 1cm in width. These are the remnants of former hollow-stemmed plants into which the sand was washed after death. Also at the top of the coal there are several large fragments of fossil wood, again filled with sand, one showing an ovoid cross-section apparently shaped like a section along a rugby ball, the deviation from the normal circular cross-section of a tree trunk indicating deformation by the mass of sediment formerly accumulated above it (Figure 10.2). The infilling sands have sufficient layers of black fragments to indicate progressive deposition within the fallen formerly hollow tree trunk. A ring of black carbon around the filling provides traces of the original bark. Above the coal seam there are several metres of thin, closely bedded sandstones, some of which bear ripple marks on the bedding.

Plate 10.2 Sands filling former hollow tree trunk in Witch Lake Cliff, St Andrews. Former circular cross-section has been deformed into an ellipse by the over-lying sediments.

As seen from the beach, the black coal seam appears to be nearly horizontal at the cliff foot. If the seam is followed to the west, the coal is seen to fall from head height to the level of the foreshore, from where it has been partly eroded by natural forces but more importantly by man, who has excavated westwards beneath the cliff for some metres. The coal seam itself is now seen to be inclined downwards beneath the town at an angle of about 2° from the horizontal. No coals were won below low-tide level as the workings would have been flooded twice daily. The seaward side of the mine tunnel has been broken through and repaired by the insertion of cement-filled sand bags lest the entire cliff face above were to collapse (Figure 10.3).

Take a few steps back from the cliff foot and look at the thick, yellow-coloured sandstones above the coal seam. On the bases of several of the overhanging layers of finely bedded and rippled sandstones, pale-coloured irregular marks can be seen. Some of these are tracks of organisms that once lived in the wet sands. The closely layered sandstones are overlain abruptly by a 3m-thick pale-coloured massive sandstone that bears few signs of internal layering. Follow this sandstone to the east (towards the Castle). While the base of the sandstone remains above the thinly bedded layers, the top of the sandstone gradually slopes down towards the Castle so that this prominent unit decreases in thickness from 3m to less than 1m before a second upward-sloping boundary

Plate 10.3 Witch Lake Cliff, St Andrews, showing the rock layers and remains of early mining activity at left.

is detected, also marking the top of the sandstone. The sloping upper surfaces mark the margins of what was an ancient river channel no more than 20m wide. The channel is filled with thin inclined layers of sand that are parallel to the western slope but thicken gradually across the channel, reaching higher and higher up the eastern margin and serving to fill the bed of the former stream. The massive sandstone on either side of the buried channel is constant in thickness. Above the thick sandstone and channel fill is a second black coal seam up to 35cm in thickness whose base sinks slightly above the buried channel. The outcrop of this second coal may be traced along the cliff, showing as discontinuous exposures appearing between patches of vegetation draping the cliffs (Figure 10.4). It is normally favoured by fulmars and other sea birds in spring, as it provides ideal nesting sites. Immediately above this coal seam is a prominent band of sandstone which has been enriched with iron oxides to produce a bright orange weathering colour.

If this coal seam is followed to the west along the cliff face, always maintaining its position above the thick massive sandstone, it will be seen that as the level of the top of the cliff falls gradually towards the west, the coal holds its level so that the two surfaces should meet on the cliff top to the east of the Martyrs' Monument. The intersection of the cliff top and the coal seam is usually obscured by vegetation. The plants may be filling a hollow along the

Plate 10.4 The two upper coal seams at Witch Lake Cliff, St Andrews. The surface of the lower thick sandstone is cut by a channel, above which a prominent coal seam hosts nesting fulmars.

crest of the face, possibly the site of an old and probably short-lived excavation into the coal outcrop.

The coal seams maintain their height above sea level along the face from the Martyrs' Monument to the Castle. This direction within a coal seam is of importance when considering possible mining activity. It marks the direction along which headings, tunnels or roadways are most normally constructed within the body of the coal seam if it is of sufficient thickness. When the coals are released to be taken to the surface, the tubs in which they are carried are most readily moved along horizontal trackways to the base of any shaft. As the rocks exposed on the foreshore and in the lowest coal seam are seen to dip landward, any mining-related activity would have been developed not at the cliff but some way below the town. If the two upper coals are traced along the cliff face to the Castle, again they may be seen to dip downwards below the Castle itself. Within the confines of the courtyard is situated the famous 'bottle dungeon'. This is said to have originated as a punishment site or a storage site for prisoners, but it is probably a coincidence that the base of the 7m deep dungeon shaft is very close to the level that the upper coal seam would be found at that position. There is no known evidence to indicate whether the coal seam had been sought to provide a source for fuel, possibly during siege conditions (D. Speirs, pers. comm., 2015). The seam would have been thin and difficult to work, had it been found.

The rocks exposed in the cliffs at St Andrews are believed to be from near the centre of a succession of rocks known as the Pittenweem Formation (see Table 1). The older rocks of this sequence form the ridges of the skerries, which are seaward and to the north of the St Andrews Aquarium towards and beyond the Bruce Embankment and extending beneath the West Sands. The base of the Pittenweem Formation lies at the seaward side of the strong sandstones here (N56.3436, W2.8000) and takes the form of a 1.5m thick mudstone that has a thin, fossiliferous limestone near its base. This is the West Sands Marine Band (I), as defined by Forsyth and Chisholm, (1977) which marks the boundary between the Pittenweem Formation and the underlying Anstruther Formation, also of sandy material to seaward of the muddy marine band.

At low tide below the foot of the cliffs at Witch Lake, a 3m thick mudstone with a thin band of limestone bearing fossil crinoids is commonly detected. Formerly referred to as the 'Encrinite Bed' (Geikie, 1902) this has now been given the name Witch Lake Marine Band (II) by the geological surveyors (Forsyth and Chisholm, 1977). Between these two marine bands there are suggestions of a third layer in the form of mudstones containing non-marine fossil ostracods within the Step Rock lake area.

On the foreshore to the southeast of St Andrews Castle (N56.3420, W2.7905) is a third marine band, dubbed the St Andrews Castle Marine Band (III), containing the fossils *Naiadites* and *Myalina* among others. The top of this marine band marks the boundary between the Pittenweem Formation and the overlying Sandy Craig Formation.

Following the river banks from the harbour into the Kinness Burn reveals that the beds of sandstone continue to dip gently southwards. Near the base a non-marine fossil band containing ostracods and fish fragments is overlain by about 45m of sandstones before the overlying mudstone containing the New Mill Marine Band (IV) is exposed on the bed of the stream about 90m below New Mill (Forsyth and Chisholm, 1977). A short way above the Mill, grey shales with fish and ostracods of a non-marine band lies below a 40cm thick shaly coal. At Cairnsmill, the southward inclined sandstones and mudstones that lie beneath sandy mudstones are probably of the Sandy Craig Formation, but as the limestones to the south belong to the Lower Limestone Formation, it is assumed that movements along the Maiden Rock Fault cut out the Pathhead Formation in this area.

East Sands to Maiden Rock, St Andrews

The first rock succession beyond the end of East Sands consists largely of southward dipping pale brown and yellow-weathering sandstones and dark grey shales, but including a 3m thick sandy mudstone seen on the foreshore at (N56.3331,

W2.7788). This includes the St Nicholas Marine Band, so named after the local farm by Forsyth and Chisholm (1977), who suggested it as the equivalent of the New Mill Marine Band (IV). Some 30m above it in the succession of rocks there is a non-marine band containing the fossil *Curvirimula*, and this may equate with the similar unnamed band above the New Mill. However, lack of exposures in the bed of the Kinness Burn makes it impossible to determine accurately the separation between the two bands. The foreshore consists of a series of prominent ridges of sandstones that have resisted erosion by wave activity, separated by excavated gulleys floored by weaker shales and mudstones. Some of these may contain thin coal horizons, but none is prominent.

Approximately 500m from East Sands, where the Maiden Rock Fault and associated fractures reach the foreshore, the rocks are strongly disturbed, forming anticlinal and synclinal folds that extend for a further 500m eastward. Below and seaward of the stranded sea stack of Maiden Rock, isolated on the outer margin of a remanent old raised beach platform (Fig. 10.5A), a marine band with the Encrinite Bed (termed II by MacGregor, 1968) occurs on the western limb of the northward plunging Maiden Rock syncline. About 50m to the east this same bed reappears on the eastern limb of the fold, and repeats its occurrence 170m to the east of the Saddle Back Anticline. At each location it is possible to locate the remains of the Encrinite Bed, the feature that led Kirk (1925) to identify it as the same horizon as the now named Witch

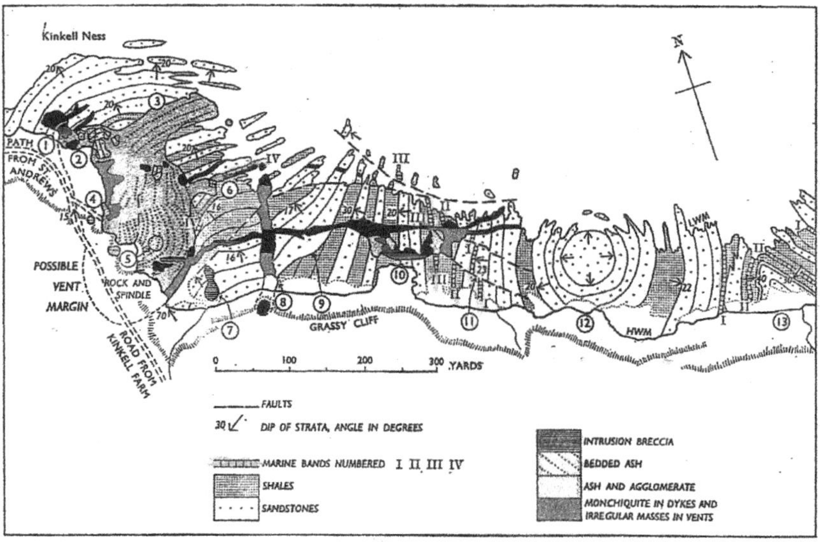

Figure 10.5A Map showing position of Marine Bands on the foreshore east of St Andrews: Rock and Spindle to Craigduff Dome (after MacGregor, 1968: adapted by Graeme Sandeman).

Lake Marine Band (II) of the cliffs at St Andrews. However, there is no sign of either of the coals a short way above the marine band as in the Witch Lake Cliffs. MacGregor (1968) provides details of a younger marine band 120m to the east, situated below a sharp gulley in the cliff line. Containing a rich fauna with *Myalina* and *Naiadites* in dark grey and black shales above thin but very fossiliferous limestones bearing *Productus, Aviculopecten* and crinoid debris, this is probably the same as the New Mill Marine Band (IV) of the Kinness Burn section.

To the east of Kinkell Ness, the swirl of the bedding is broken through by the prominent layering of the largely vertical ashes lining the margins of the vent of the Rock and Spindle volcano (Fig. 10.5B). On its north-eastern margin, the New Mill Marine Band (IV) reappears, and some 300m to the east, beyond a series of alternating sandstones and shales outcropping east of the Dome, the faulted First, Second and Third Marine Bands are recognizable. Their outcrops may be traced across the wave-cut platform about 150m from the centre of the Craigduff Dome. Both the First and Second Marine Bands are exposed on the limbs of the southward-plunging Craigduff syncline. The main sandstone between the two Bands is 12m on the western limb, but on the eastern limb it is much thinner at 3m. This variation is similar to that seen above Marine Band II at Witch Lake, where erosion and fill of a former stream channel were postulated. Here however, the base of the strongly current-bedded sandstone, which

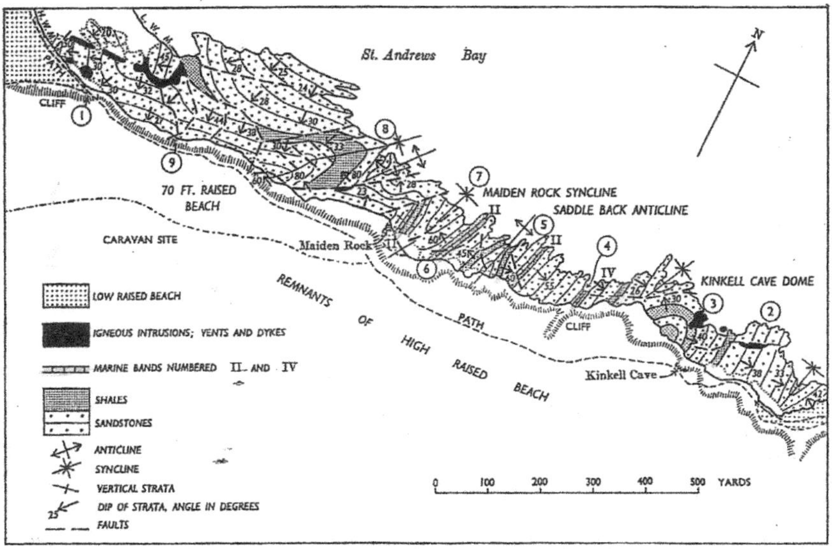

Figure 10.5B Map showing position of Marine Bands on the foreshore east of St Andrews: between East Sands and Maiden Rock (after MacGregor, 1968: adapted by Graeme Sandeman).

contains many large fragments of tree roots and trunk pieces, can be seen to cut progressively into the underlying shales. A coal seam 30cm thick is exposed almost immediately below a yellow-weathering limestone of Marine Band III (the St Andrews Castle Marine Band), on the eastern side of the volcanic vents to the west of the Dome. Where in contact with the volcanic vent material, the coal has been altered by heating and has become hardened, a feature that is repeated in many places throughout the East Neuk.

Following along the coastal path, the next major feature to be encountered is Kittock's Den where, by the mouth of the Den, the Witch Lake Marine Band (II) is again exposed in the cliffs above a series of thick sandstones. Further east towards Buddo Ness there is a 3.5m-thick mudstone with a marine fauna containing crinoids, bivalves and *Lingula* enabling it to be equated with West Sands Marine Band (I) at St Andrews. As is common in the marine bands, there are also ironstone nodules within this mudstone. In the folded succession below this horizon and extending to the Buddo Ness Fault, the foreshore sandstones include two thin dolomitic limestones (of calcium-magnesium carbonate), which contain a distinctive non-marine fauna. The limestones are marked on the Geological Survey 1:50,000 scale maps, Sheet 41(S) and 49(S) and may be traced for 6km eastward as far as Kingsbarns, where a thin coal seam is present at the foot of the dunes below the car park at the end of Sea Road. These rocks were deposited before those of the West Sands Marine Band (I) and are considered to be the upper part of the Anstruther Formation (Forsyth and Chisholm, 1977). To the north of the Craig Hartle Fault the 70m succession of rocks is older than those higher up the foreshore and the Craig Hartle North Marine Band provides a useful marker within the Anstruther Formation. Along this coastline no coals, even the thin coal seams seen at St Andrews, have been recorded, although there are traces of plants to be found in many of the sandstones and siltstones.

It is not proposed to examine every rock exposure and marine or non-marine bands along the Fife coast, but attention will be drawn to some of the more interesting features that can be readily identified.

The western side of the Boarhills inlet about 500m southeast of Craig Hartle reveals the western limb of the Craig Hartle syncline where a succession, about 50m thick, includes several thin layers of coal and also thin dolomites. Fossil tree trunks are preserved in sandstones near the base of these exposures. The Craig Hartle South Marine Band, containing crinoid ossicles and bivalve molluscs, occurs near the top of the west limb section and the Kenly Mouth Marine Band is exposed at (N56.3190, W2.6793). All of these rocks belong to the Anstruther Formation, which is normally lacking in coals. Both limbs of the Babbit Ness Anticline show alternating beds of sandstone, siltstone, mudstone and seatearth with dolomite and ironstone horizons, but again no coals are preserved.

The coastline swings towards the southeast beyond Babbit Ness and the gently inclined westward dipping rock ridges trend parallel to the line of the coast. The lower units, exposed to seaward, are of sandstones, mudstones and dolomites but the landward sections are more varied, with the addition of siltstones and shales bearing a distinctive non-marine fauna. Some of the exposed bedding surfaces exhibit parallel lines of small depressions, tracks of animals, mainly arthropods, which lived in the shallow waters. A marine dolomitic limestone is present northwest of Kingsbarns Harbour. Throughout much of this succession plant fragments are scattered on the bedding planes, but no coal seams are preserved. The rocks here are from low in the Anstruther Formation.

The public car park beside the Cambo Sands (N56.3034, W2.6450) provides easy access to a block of sediments from the Upper Old Red Sandstone that has been uplifted between the Kingsbarns Harbour Fault and the Cambo Ness Fault. Near Cambo Ness they are overlain by the sandstones and mudstones of the Fife Ness Formation where a thin coal seam is present in the cliffs. The black coal seam, which is no more than a few centimetres thick, sits above a very pale sandy seatearth and below a yellow-brown bed of limestone.

To the south of the Cambo Ness Fault between Cambo and Randerston eleven thin limestone beds trend southwest to northeast across 2km of foreshore. Kirkby (1880) and Geikie (1902) likened intervening sandstones to those of the Anstruther–Pathhead Formations, but Forsyth and Chisholm (1968) suggested that there are closer similarities with material from the lower part of the definitive modern borehole record from Anstruther. Although the marine limestones are varied and of interest in themselves, no further reference will be made to them here, as no coal seams are associated.

To the east of the Randerston Castle Fault, (N56.2897, W2.6204) the rocks are also thought to be from the Anstruther Formation but from a level below the base of the material recovered from the Anstruther borehole. They are therefore older than those seen between the Cambo Ness and Randerston Faults. Ten well-developed cycles of sedimentation are recognized. Seatearths, very thin coal seams and occasional dolomitic layers occur throughout, and several marine bands are present within the sequence, but the fossils suggest that they were earlier in age and have not been linked with the succession at St Andrews.

More supposed Upper Old Red Sandstone rocks appear in the Tullibothy Craigs before being conformably overlain in a continuous section by thick, white-coloured fine-grained sandstones, grey and red mudstones and grey siltstones of the Fife Ness Formation, probably the oldest Carboniferous rocks in Scotland. The 20m thick Fife Ness Sandstone, one of the thickest layers, occurs at Fife Ness but there are no known seatearths or coals in this formation.

The southward downthrow of the Dane's Dyke Fault brings younger rocks to the surface to the south, as do smaller faults towards Kilminning Castle. North of the Castle is the Goats Marine Band, a dolomitic limestone with many cephalopod fossils. The *Carbonicola* rich sandstone about 15m above it marks the position of the Kilminning Castle Mussel Band. These two marker horizons are believed to be equivalents of the Anstruther Wester Marine Band and the Kilrenny Mill Mussel Band a short way to the southwest, both of which pre-date the rocks exposed at St Andrews.

One kilometre to the southwest, the eastern side of a shallow, faulted dome meets the coast at the eastern limit of Roome Bay (N56.2612, W2.6131) on the eastern outskirts of Crail, where a thin coal seam varying below 0.65cm in thickness is overlain by a dolomitic limestone and sandstones. The coal was formerly worked from the foreshore and ancient workings were detected in the Pinkerton housing development in Roome Bay Crescent in the mid-1980s. The same coal is believed to have been present in exposures by the former nunnery and church wall excavated in the 1930s. A coin from the 1550s was found among the mining debris, suggesting that the workings may have been active during the middle of the sixteenth century. The coin may have been a votive offering left on closure of the mine. The western limb of the structure returns to the coast below the cliff-top footpath east of Crail Harbour. The location of the harbour, like most of similar facilities in the East Neuk, is principally defined by the presence of a significant fault, in this case the Crail Harbour Fault which, like most local faults, has a downthrow to the southwest.

To the southwest of the harbour the youngest beds observed are principally of a thick marine mudstone above thin limestones with ironstone nodules. This is the Crail Harbour Marine Band (Forsyth and Chisholm, 1977), identified from its brachiopod, mollusc and crinoid fossils as the possible Witch Lake Marine Band (II) equivalent. The cliff foot is of a range of thin sandstones overlain by a lilac-coloured muddy seatearth beneath a thin coal seam which is, in turn, overlain by a prominent sandstone no more than 1m in thickness (Figure 10.6).

Several fossil tree stumps more than 30cm high and of greater diameter stand rooted above their seatearth (Plates 10.7A and 10.7B), giving a clear indication of the scale of the plants forming the forests that grew here in the distant past. ***Photograph them (with a scale) by all means, but please do not use hammers on them.*** Many of the bedding planes here reveal sets of pairs of footprints of both large and small arthropods that foraged on the forest floor. A slab of rock with excellent displays of these features recently slipped from the cliff, but remains visible amid the foreshore boulders. Try to estimate how many animals wandered across this area and think of their sizes. Did any leave tail trails? Some of the arthropods may have been more than a metre in length.

A short way southwest of the Coves (N56.2421, W2.6468) is a succession of thick sandstones with cements partly derived from circulating groundwaters containing both calcium and magnesium in varying proportions. Such changes may have occurred when the saline tropical waters circulated within the soils

Plate 10.6 Thin coal seam in the cliff west of Crail Harbour below prominent sandstone.

Plate 10.7A Fossil tree trunk in growth position in sediments west of Crail Harbour (scale bar 30cm).

Plate 10.7B Fossil tree trunk standing up through, and buried by, layered sandstones and shales west of Crail Harbour (scale bar 30cm).

and underlying rocks soon after the start of a period of flooding. Reddening of the rocks suggests that some of the ironstones were also being altered by oxidation at this time, as in the formation of dolomitic bands elsewhere in the succession. One sandstone, over 20m thick, lies above the non-marine dolomitic Caiplie *Lingula* Band. Many of the sandstones bear plant fossils, including trunks of *Calamites* (see Figure 4.2), in rocks deposited soon after the start of a period of flooding. At least one horizon appears to be partly of coal and partly of limestone, bearing structures suggesting that the limestone is, or was, chemically replacing the coal, producing pseudofossils as it did so.

The lower 300m of sandstones of the Anstruther Formation lead to the shoreline at Cellardyke. There are no coals in this part of the succession, but it has been suggested by Forsyth and Chisholm (1977) that the Kilrenny Mill Mussel Band lies within the thick sandstones near the entrance to Cellardyke Harbour. The thick sandstones on the foreshore to the east of Anstruther Harbour have been correlated with the top of the Anstruther Borehole, referred to by Forsyth and Chisholm (1977), which was drilled from the old pier at Anstruther Wester close to the axis of the Anstruther Anticline.

At Anstruther Wester the essentially cyclic nature of the deposits of the Anstruther Formation is well displayed. The cycles are mainly those of upward coarsening of the sediments with repetitions passing from thin limestones (A) often absent, into mudstones (B) shales (C), siltstones (D), and sandstones (E) with seatearths (F) and coals, above which is a surface marking a relatively rapid influx of the sea before the sequence of deposition is repeated. It is a pattern that runs through from A,B,C,D,E and F and immediately repeats with A,B,C,D,E, and F, i.e., a pattern indicating forward growth of a delta, with no sign of a gradual encroachment of the sea to flood over the surface, which would have given a reversal of the pattern of change to A,B,C,D,E,F,E,D,C,B,A. Greensmith (1965) interpreted this succession as one of fluviodeltaic sedimentation, which repeatedly built forward into shallow inter-distributary embayments where the salinity of the waters was variable but commonly low. This was supported by later workers such as Belt (1975), Fielding et al (1988) and Burn (1990). The significance of the flooding surfaces which could be recognized over wide areas of deposition (Weibel, 1996), was adopted by Kassi *et al.* (1998, 2004) in more recent interpretations of the sequences. The influxes of the sea were evidently swift and catastrophic to plant and animal life in the coastal areas. In most cases the thin, drowned palaeosols referred to as coals are described as 'duff' coals, which are coals, coaly shales or coaly siltstones containing much clay and silt and, in consequence, do not burn well.

The succession of rocks along the foreshore from Anstruther Wester starting at the west pier is recorded in bed-by-bed detail in Geikie (1902) who quotes the results of many years of study by Kirkby, who had died by the time of publication. Kirkby's carefully numbered record measured no fewer than 761 beds of rock, starting at the well-defined St Monance Brecciated Limestone, outcropping below Pathhead near St Monans. He recognized as many as 44 coal horizons, but few of these exceeded 0.3m in thickness, and even fewer were thick enough to have been worked in the past.

The westward dipping rocks on the foreshore at Anstruther Wester are mainly of grey mudstones and siltstones with a few very thin coal layers. Towards Billow Ness and starting from 10m below the base of the 12m thick Johnny Dow's Pulpit sandstone, Kassi *et al.* (1998) recorded a sandstone more than 3m thick being overlain by three leaves of coal with two internal shale beds and another at the base. The thick sandstone passes up into about 7m of very varied sediments, including ten upward coarsening beds of siltstone culminating in a 2m shale that contains many ironstone nodules. This layer is terminated by a root horizon, above which lies a second coal assemblage with intervening sandstone and shale layers. The lowest leaf of this coal is 1m in thickness. Directly above the coal are a thin dolomitic limestone and a series of mussel bands in shales. These are in turn overlain by two 3m thick current-bedded sandstones, before passing upwards into thin coals and shales. The terminal thin coals of this cycle of deposition lie immediately below the Billow Ness Marine Band. In all, a total of ten sediment cycles are detected along this shore; those discussed in the Billow Ness area are from cycles 4 to 7.

Continuing westwards along the coast below Chain Road leads to a faulted outcrop of the grey siltstones and shales, which culminate in the Chain Road Marine Band. The rocks of the foreshore are well exposed, and no more than 200m to westward, the base of Cuniger Rock (N56.2150, W2.7165) rests on the Cuniger Rock Marine Band that marks the top of the Anstruther Formation and the base of the overlying Pittenweem Formation. It is equated with the West Sands Marine Band of St Andrews.

Pittenweem Formation

The rocks of the Pittenweem Formation are mainly of alternating mudstones, shales and thin sandstones with thin seatearths and coals that dip up to 40° to the northwest. There are three marine bands in the Pittenweem Formation at this locality. The Cuniger Marine Band marks the base, the Kirklatch Marine Band in mid-sequence, and towards the top is the Pittenweem Marine Band, a richly fossiliferous 6m-thick mudstone bed with carbonate horizons. Some 10m above it is the Pittenweem Harbour *Lingula* Band, immediately below a 0.65m coal seam that Kirkby believed had been worked in the past. The base of the Pittenweem Formation follows directly upon the Anstruther Formation sediments without any break. However, the top of the Pittenweem Formation is separated from the succeeding Sandy Craig Formation by the northwest-trending Pittenweem Fault.

Sandy Craig Formation

Only the upper part of the Sandy Craig Formation is seen on the Fife shoreline because the base is faulted out by the Pittenweem Fault. Although, according to Forsyth and Chisholm (1977), two further faults cut out other parts of the succession, at least 550m of the Formation remains exposed to view. The oldest rocks of this Formation outcrop between the harbour and the Pittenweem fault. Most are grey-coloured mudstones, siltstones and sandstone with pale dolomites near the base. Several thin layers with fine gravel are of possible volcanic origin, but no coals have been recorded in the lowest 100m exposed. This part of the succession extends as far as the harbour entrance.

Thick sandstones, some of which are coarse-grained and poorly bedded, lie between root-bearing mudstones. Towards the top of the succession, red- and yellow-spotted mudstones occur in a pebbly layer with lenses of coarse carbonate conglomerates. Neither coals nor marine bands are known from this part of the column.

The Boat Harbour Fault in the northern part of the basin trends north-north-eastwards into the Sandy Craig Fault. The 120m thick north-westward dipping sequence is mainly of mudstones, some grey and root-bearing, others showing red and yellow mottling, and there are dolomite and siderite concretions within the layered rocks. A coarse sandstone and a carbonate conglomerate both occur in the central part of the succession. To the north of the Sandy Craig Fault, about 140m of the uppermost thick sandstones of the Sandy Craig Formation are well exposed in the foreshore leading to the Pittenweem Bathing Pool. This sandstone unit is coarse-grained and contains small quartz pebbles. It rests upon a succession of red mudstones, often hosting trace fossils and including both a dolomite horizon and a 2m-thick lens of cross-stratified carbonate conglomerate, indicating that both limestone and dolomites were being eroded relatively close at hand.

Beyond the headland by the former Pittenweem Bathing Pool, the Pathhead Formation is exposed for about 700m along the inner foreshore as far as Pathhead, the youngest division of the former Calciferous Sandstone Measures. Well-described in the guide book of MacGregor (1968), the succession continues to be mainly of sandstones with interbedded mudrocks, thin coals and three prominent limestone horizons surrounded by marine shales. The essentially cyclic patterns of deposition are retained, but now the Lower and Upper Ardross Limestone, the St Monance White Limestone and several substantial sandstones form prominent ridges crossing the foreshore. Marine mudrocks dominate the sediments. Those overlying the thin coal seams vary greatly in thickness from a few centimetres to several metres, a feature that Greensmith (1965), and others have attributed to the different rates of subsidence within the ancient delta. The general lack of erosional features within the successions overlying the coal seams suggests that the influxes of the sea were rapid, so that there was not enough time for the local streams to cut channels into the newly flooded surfaces. Later in the successions, some of the sandstones exhibit erosional bases pointing to their origins in conditions of active stream erosion, probably associated with the falls of sea levels providing progressively lower base levels towards which the streams drained. Other sandstones show current-bedding, and cross-sections of the ripples indicate that the currents responsible for their formation changed their flow directions during deposition as though developed under the influence of reversing tidal flows.

About 5m above the base, a low ridge of oil shale is readily identified from its patchy yellow coloration. This, the only such shale known to be exposed on the shoreline, is similar in texture to earlier oil shales formerly worked for several years in inland quarries, notably near Kilrenny. The uppermost cycle of deposition in the Pathhead Formation, that containing the St Monance White Limestone, is regarded as that of a river-dominated but marine-influenced deltaic environment leading into a shallow offshore marine environment. Minor cycles of deposition marking the passage of individual storms are preserved locally within the pattern of major cyclic sedimentation controlled by the global rises and falls of sea level.

The Lower Limestone Formation succession begins at the path to the foreshore beside the nursery at Pathhead. The rocks forming the St Monans–Pittenweem coalfield lie to the north and west of this point, forming a block of deposits that terminate against the Ardross Fault, a major fracture line extending 9km from Elie to Kilrenny. To the north of this structural feature lie further deposits of the Pathhead Formation, which include the easternmost substantial coastal coal sequences initially identified by Landale (1835).

Rather than continuing to discuss the stratigraphic sequence towards the harbour at St Monans to Newark and Elie, it would be useful at this juncture to trace the coals inland away from the coast.

The inland coals of the East Neuk

In the creation of the First Statistical Account of Scotland, the local ministers were required to submit brief accounts of life in their parishes. In most cases they concentrated on the population, the nature of the local agriculture, and the varieties of work undertaken in the parishes before considering any relevant general or historical knowledge of the area. The amount of information relating to the coal industry varied according to the interests of the writers. Some provided detail of individual workings but, alas, the majority of the accounts made scant reference to the coals and the colliers working in the industry. It was as though the industry and its workers were thought unimportant and, following local opinion, members of mining communities were considered as in some way sub-human, working below ground as many did for most of their lives.

In the First Statistical Account of Crail, Bell (1790) drew attention to the fact that thin coals had been worked in fields to the east and west of the parish and that, at that time, the former workings were still visible. To the east of the village only the south-eastward dipping crop coals had been worked, always at shallow level, precluding the need for the introduction of pumping to remove any groundwaters. There was no indication that machinery had been used to extract the coals. Bell also referred to the discovery of the gold coin from the reign of Queen Mary, dated 1553, found on the shore near the former coal workings on the east side of Roome Bay. This may have been a votive offering marking the closure of the workings.

The Second Statistical Account (1845) refers to westward-dipping coals west of the harbour. These had been worked quite extensively some years previously at Kirkmay Farm, west of Crail, where Robert Inglis had used a steam-powered engine for a short period to remove the groundwaters. This working was short-lived, supplying coals for lime burning and for a brick and tile works. The uppermost of the three seams consisted of 0.7m of coal (in two leaves), with a 0.15m-thick band of shale above a fireclay pavement. Excavations in 1933–34 revealed shafts and coal wastes at the eastern and western ends of the nunnery (Dott, 1944). A similar coal 0.9m thick was discovered in the churchyard during extensions to the graveyard. At various times coals from further up the Firth of Forth were brought into the harbour by small coastal trading vessels and distributed locally by farm carts.

The rocks on the foreshore at Kilrenny are dominantly of sandstones, mudstones and shales with occasional bituminous limestones and thin coals. On the shoreline an important 2m-thick limestone rests above a coal seam 0.5m thick. The association of the two materials, each independently of marginal value, enabled both to be worked profitably in tandem inland at Cornceres. As usual the limestone was worked in an open quarry, and the coal in a pit nearby. The coal thickened to 0.8m in a little over half a kilometre from the shore.

During the 1860s the shallow (5.5m) No. 1 Kilrenny Shale Mine is believed to have worked the Pitcorthie Coal near Graham Yooll's oil shale works. Yooll was also operating the small Balcormo Pit south of the Kellie Castle grounds at that time. According to Sparling (2005) by 1873 Yooll was managing small shale, coal and ironstone workings near West Pitcorthie, 2m north of Kilrenny. By 1875 this employed 12 men above ground and five below. The shales yielded a prolific fauna of fossil fishes (Traquair, 1901). The nearby Lady Erskine Pit is known to have been sunk to a depth of 24.7m but appears to have ceased working before 1878.

The interest in oil shales was kindled principally through the work of 'Paraffin' Young who, in 1851, introduced the use of coal into gas-making technology by heating cannel coal in an enclosed vessel and condensing the oil vapour, from which several valuable mineral-oils were produced. As the cannel coals became scarcer attention moved to the oil shales, which were more common in Central Fife and the Lothians. In essence, the oil distilled from shales at 900° was reheated to yield ammonia and hydrocarbon gases. The Broxburn Shales yielded 24–30 gallons of crude oil per ton. None of the oil shales of the East Neuk produced at this level.

Temporary exposures in limestones and shales about 2km east of the West Pitcorthie workings at Thirdpart (N56.2518, W2.6630) yielded fish fossils similar to those at Pitcorthie (Wattison, 1962). About 2km to the north-west at Ribbonfield, boreholes reaching 95.7m from the surface again revealed the typical Anstruther Formation successions of sediment, with fish remains and bivalves in non-marine shales associated with thin limestones and coal seams. A series of now inaccessible limestone quarries north of Ribbonfield yielded workable shelly limestone at least 2m in thickness, and although several thin coal seams were recognized, none was sufficiently thick to warrant working.

About 1km east of Carnbee lies the village of Balmonth where stream sections show the sediments to be mainly fine-grained shales and mudstones with three thin dolomitic limestones, the thickest no more than 0.06m, and there are several thin coals up to 0.4m thick. These clay-rich rocks lack fossil evidence to permit precise identification beyond 'probable Pittenweem Formation' (Forsyth and Chisholm, 1977).

To the south at Ovenstone, (N56.2378, W2.7565) the dominantly muddy rocks contain marine fossils, and blocks of crinoidal limestone confirm their marine origin, possibly at a level above the Pittenweem Marine Band. There is a 0.4m-thick coal seam that has been greatly altered, even coked, by intrusive igneous rocks.

At Grangemuir, where there are few surface exposures, BGS records show the presence of 20 boreholes drilled between 1897 and 1900. The logs show the typical sandstone, siltstone, mudstone successions of sediments with a few thin limestones and rather more coal seams, one of which was noted as 0.6m in thickness. The fossils preserved contained no index markers and the sequences could be part of the Pittenweem Formation, the Anstruther Formation, the Pathhead Formation or even the Lower Limestone Formation (Forsyth and Chisholm, 1977, p.56). The Walter Pit was worked briefly, producing a rather soft coal that burned well and produced little ash. When first opened, the pit faced problems with water, but the introduction of an improved steam engine surmounted that problem. An accident that led to James Croll, the engine man, having part of his foot amputated, led to the closure of the workings in March 1903.

The Kellie Coals

One of the most important and nearly continuous early groups of coals to have been worked in the interior of the East Neuk are what Landale (1835) referred to as the 'Kellie Coals' of the *Fore Coal* and *Back Coal*. The principal seams were worked discontinuously over a distance of 8km from Balcaskie, near Kellie Castle (N56.2369, W2.7763) in the south to possibly as far as Bridgetown in the north, midway between Stravithie and Lathockar. During his long career in the coal exploration business, Landale visited many sites in these two seams and was less than enthusiastic about them, making somewhat disparaging remarks that they were 'important to the geologist but are otherwise of comparatively little value' (Landale, 1837). However, in later works he provided a more thoughtful appreciation of the two coals. Much of what follows is based on his contributions of 1854 and 1868. The Fore Coal and the Back Coal, both of which dip towards the northwest at low angles, lie within the Sandy Craig Formation.

Between Kellie Castle and Arncroach four uniformly dipping coal seams are known to be present. Only two of these were worked. The lower *Fore Coal* ranges from 1.4m to 2.1m in thickness and incorporates a 0.05m shale horizon 0.6m above its base, and a second similar 0.08m thick layer occurs near the roof, which is of a hard 0.9m-thick shelly limestone. The *Back Coal* is 42m above the Fore Coal. It is of near-constant thickness at 2.7m thick, but incorporates three shale horizons equally spaced through the bed. This seam has a 2.7m thick roof of a hard, black bituminous mudstone formerly incorrectly referred to locally as 'slate' and which is resistant to weathering.

The undated map, located by Forbes (1955) probably originating from one of Landale's reports of 1854 or 1868, shows the coalfield to have consisted of three parts separated by two substantial west-north-west to east-south-east trending faults (Fig. 10.8). Both show southward downthrows of the succession. The Balcormo Fault, the southernmost of the two, has a downthrow of

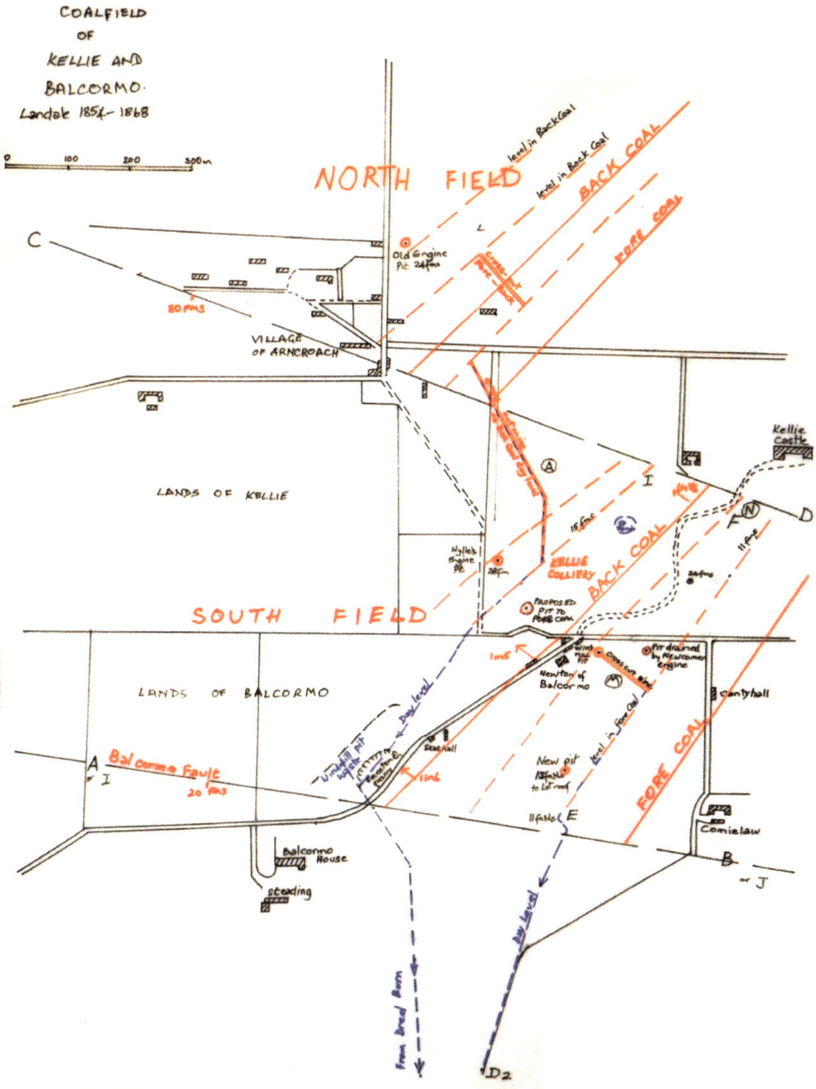

Figure 10.8 Map of the Kellie Castle–Arncroach Coalfield (after Landale, 1868).

37m but the northern Arncroach Fault has an estimated downthrow of 146m. The South Field lies between these two faults and the North Field, in which there had been no working for many years before the 1854 report was prepared, is to the north of the Arncroach Fault. As the coals were worked progressively northwards towards Kellie Law, they became hardened and 'blind' as a result of baking by sills of igneous material and exhibit many small faults on approach to the higher ground. This working of the Fore Coal was by the Earl of Kellie in the early part of the eighteenth century and appears to have been achieved without the aid of a pumping engine of any sort.

In 1877 the estate and its minerals passed to Sir John Anstruther, who proceeded to drive a cross-cut stone mine (CD) from the Fore Coal to the Back Coal and then worked by room and stoop northwards at the level of the Fore Coal as far as practicable under the area (DEF). The waste from this segment of the field caught fire, resulting in the residual pillars in the mine being fully destroyed. The quality of the coal improved and also hardened more towards Arncroach and the adjacent lands. Later the coal became the property of Sir Robert Anstruther, Bt., of Balcaskie. A 44m-deep engine pit was constructed at Arncroach and the coals below the north-western segment were removed for as far north as they remained in good condition.

The Back Coal often 'sat' (collapsed) and the pillars crushed into the soft pavement beneath, causing depressions on the surface and this, with fires in the underlying Fore Coal, allowed a considerable influx of water into the seams so that the cross-cut gallery had to be dammed at (B) and the North Field wastes became virtually filled with water.

The workings of the South Field were closely linked with the Balcormo estate and were operated jointly as a single enterprise, Sir Ralph Anstruther and Mr Smart representing each estate, which received rents of £70 annually. In 1854 Landale wrote of having had access to a plan dated 1766 referring to workings then 'beyond the memory of any...living'. That implies that the early mines operating hereabouts were coming to an end in the 1730s or 1740s, but we have no clear indication of when their workings began – presumably a decade or more before closure.

In the South Field the Back Coal was drained by a separate day level leading from the lands of Balcormo south towards Abercrombie Mill, beyond which it flowed into the Dreel Burn and away to the coast. This day level met the coal at (P) (Fig. 10.9) at a depth of 27m and the coals were extracted from (P) to (P2), a short way north of the Balcormo march wall and extended to the line of the Arncroach Fault at (I). It is from this day level that a 300m-long cross-cut stone mine was established leading north-westwards and passing below the intersection of the Fore Coal outcrop with the fault line near (J). The

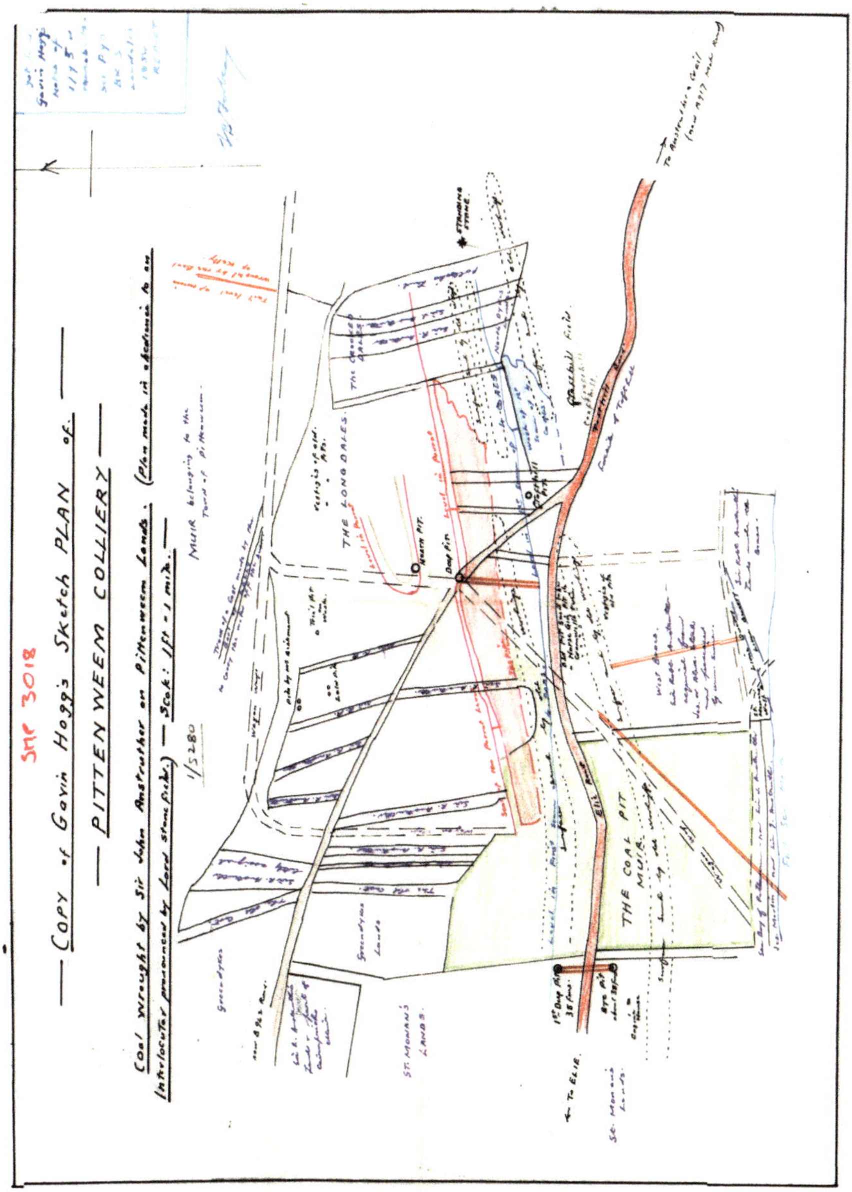

Figure 10.9 Hogg's map of the Pittenweem Coalfield, 1785.

cross-cut mine led to a 'dip-head' level following the direction of strike of the Fore Coal towards (K) 500m north of the fault. About 200m north of the fault, a further 90m long cross-cut mine towards the north-west (CD) intersected the Back Coal and from there the mine was established to work in the North Field around Arncroach. An engine pit beside the road north from the village provided a level for working the Back Coal.

At this development entirely within the lands of Kellie, the Laird of Balcormo objected and plugged the western day level as it entered the Balcormo lands. The water built up in the day level of the Back Coal and eventually escaped at the surface in the orchard at (R). Undeterred, the Earl continued mining in the north towards Arncroach, using alternative existing day levels.

Landale referred to an attempt to work the coals by means of a windmill to pump the waters from a depth that he conjectured to be between 36m and 38m. The location of this pit is probably marked as the Kellie Colliery on the surviving plan near (B) at the southern end of the early cross-cut mine near the centre of the South Field. There was concern over working this segment of the field for fear of holing through into older mine wastes that would have been filled with water, the release of which could have been violent and destructive.

The final phase of working occurred in 1847 when Wylie put down a 51m-deep engine pit in the Kellie coals beside the road south-east of Arncroach. The new pumping system carrying the waters to the surface was operated for 24 hours a day in winter, reducing to 18 hours during the summer. By this time, the underground day levels had largely collapsed or become blocked, and Landale suggested that any waters discharged into them would have rapidly found their way back into coals awaiting exploitation. When pumping began, the volumes of water to be removed were very great and the early stone mine (A), leading to the North Field was dammed and plugged. The dam at (A) held securely for about a year before the waters found an alternative route to follow and returned to the workings. The acidic waters attacked the boiler and pumps, which were expensive to replace. In winter the furnace alone required 6 tons of coal daily, which was close to the output of the 13 colliers and one oncost man.

The Back Coal worked from the Wylie pit contained no fewer than eleven layers of material, five thin shales 0.07–0.21m in thickness and six forming coal seams. The uppermost thin coal and shale horizons immediately below the roof were too thin to work. The third layer was of high-quality cherry coal but contained iron pyrite nodules. The fifth layer, the *Middle Coal*, was of poor quality with thin films of shale. It varied in thickness between 0.46 and 0.66m, being thickest towards Balcormo. A seventh layer, the 'Little Bottom' was very variable with thin layers of iron pyrites. Overall only 1.12m of usable coal was present in this seam, and that often of poor quality.

The coals dipped west-north-westwards. They were most steeply inclined at 1 in 3.5 (about 20°) in the east near Kellie Castle, and sloped at lower gradients of 1 in 5 and 1 in 6 (11° to 9°) towards the west, near the Balcormo fault. There were few rooms left to work to the north of the Balcormo March, and in 1854 it was foreseen that the main future production was likely to be from removal of pillars until a new pit might be constructed. The method of working the

Back Coal here had been essentially a room and stoop operation, but with long pillars 2.7m thick and rooms up to 4.6m across. Landale (1837) noted that in general as pillars of coal left to support the roof of former workings were removed, (the process known as 'harrying'), the response of the roof materials differed according to their composition. When the roof was of shale or slate/ clay, it bent like a piece of flexible leather without fracture. Sandstone also bent readily, but developed fine, closely spaced narrow fractures. One 40 square metre unsupported limestone roof stood for a considerable time before gradually sinking down to the pavement 1.5m below without fracturing.

It had not been a very profitable enterprise; the royalty never having exceeded the fixed rent of £100 per year. The fitting out of the workings, damming and cleaning out the mines was estimated to have cost £1300, and the engine, pumps and rails would have required a further £700, giving a total cost of £2000, which was unlikely to have been recovered from the coal sales.

The two Kellie Coals were traced north from the eastern side of Arncroach, but as they approached Kellie Law, the seams became steeply tilted and altered. They lost much of their moisture and volatiles content and became harder, changing from the familiar bituminous coals, eventually becoming blind as a result of baking by heat from the volcanic ashes and igneous rocks forming the Law. There was no record of the coals being sufficiently raised in rank above the bituminous level to develop the highly valued anthracites sought by industry. The south-eastern end of the Lathones Fault cuts through the hill mass and displaces the coals down towards the north.

North of Kellie Law traces of both of the former coal workings may still be detected following north-north-easterly outcrop paths towards two small volcanic necks at Over Carnbee. The Lochty area, a short way north of the village, is one where the main Kellie coals should be dipping westwards in an area of valuable reserves. However, Forsyth and Chisholm (1977) reported that, subsequent to the work of Landale and almost too late for Geikie to include in his account of the area, exploratory boreholes were drilled at Lochty in 1890–1900 where, to great acclaim in the local press, a 2.25m thick coal seam was located in one borehole. Kirkby (in Geikie, 1902) suggested both the Back and Fore Coals should be present 67m and 104m below the Lochty Marine Band (now termed the West Braes Marine Band). However, no continuity of these coals was found. At least seven boreholes were drilled in the Lochty area, and a summary of the findings is presented in a compilation stratigraphic column in Forsyth and Chisholm (1977, Fig. 7). In the absence of usable coals, they suggested that the sedimentary rocks could be correlated with part of the Sandy Craig Formation rather than the Pittenweem Formation, as had been previously suggested.

West of Kingsmuir House (N56.2667, W2.7507) several faults are encountered. One pair of west-north-west trending parallel faults provides a small horst structure, the central area having been uplifted, and the coal outcrops are displaced a kilometre to the west. The outcrop pattern of the two coal seams is resumed for a further 700m before they encounter the substantial west-north-west trending Lathockar Fault. There are records of former coal workings almost continuously from Arncroach to the Lathockar Fault, but beyond the line of this fracture there are no more than a few remnants of very old workings, some of which Landale (1854) interpreted as including the Kellie Coals.

Between the Lathockar fault and the Cameron Burn, the BGS 1:10,000 scale map published in 1855 showed a surface depression leading into Brigton Den, possibly marking the position of a glacial meltwater channel trending north-north-eastwards. On its eastern slopes a 1.2m-thick coal seam dipping westward at 1 in 6 (9°) was marked below a bed of yellow sandstones but above white sandstones. This is believed to be the Back Coal of Kellie complete with its characteristic roof composed of a metre or so of black bituminous shale. As the seam thinned northwards, the coal quality decreased. At Gilmerton there was no more than 0.5m of workable coal dipping to the west, and which Landale considered not worth developing. In the dip direction the locations of several pit shafts are plotted around (N56.2867, W2.7979) north of Kinaldy Farm and in the fields east of Gilmerton. A 1.2m-thick coal seam is present 25m below the surface. In a distance of 400m to the north-west, the coals were shown by Landale (1837) as having changed their dip to a northward direction, but that was not confirmed on the later BGS map.

Landale interviewed an unnamed former miner who indicated that above the Back Coal a further seam 0.8m in thickness was worked from several pits towards Gilmerton Cottage. The root of this coal was a thin sclit (shaly coal) layer above which lay a quarried freestone. The miner had worked in 14.5m–16.5m deep pits near the former smithy where the product was of cherry and rough coal, but both of these collapsed when weathered and could not be sold unless freshly worked, so the pits were abandoned.

Approximately 250m east of Brigton House, a 0.6m coal seam formerly outcropped immediately above a thick white quarried sandstone dipping west at about 1 in 5 (11°). The seam, which had a 0.6m-thick shelly limestone roof, contained both cherry and rough coal which, as above, did not resist weathering and, proving difficult to sell to all but the local smithy, was abandoned. Both Landale and Geikie thought the coal to be the equivalent of the Fore Coal of Kellie. It may be present at depth below the grounds of Gilmerton Cottage, but the quality was predicted to be inadequate to make its working economic. Forsyth and Chisholm (1977) noted that the fauna in a thin limestone within the Brigton Marine Band and Craig Hartle Marine Band suggested that the Gilmerton–Brigton deposits might actually be older than the Kellie Coals, possibly pre-dating the Sandy Craig Formation. If this is the case, then the movement on the Lathockar Fault is much greater than hitherto suspected.

A dotted line on the modern BGS map indicates the positions of several old pits along a north–south line west of the Dunino Parish boundary marking a formerly worked coal seam some 300m to the east of the Fore Coal outcrop. This coal, draining northwards by a day level leading to Cameron Burn, may be the equivalent of the *Thin Coal* noted to the east of Kellie Castle.

The Kellie Coals between Balcaskie and Gilmerton dip westwards, and it was logical to seek continuations of the two main seams in that direction from their outcrops. The records of exploratory borings undertaken in 1890–1900 and in 1936–1938 in the area of Lochty are summarized in Forsyth and Chisholm (1977, Fig. 7). The older boreholes showed the Back Coal as 2–2.25m thick with several shale partings and a black shale roof, but in the later exploration this coal was no more than a few centimetres thick in Lochty No. 3 (N56.2608, W2.7668) and appeared to be absent in Lochty No. 2 (N56.2578, W2.7726). This suggests that the seams become thinner and less likely to be workable to the west. It is quite possible that neither of the coals discovered in these boreholes was actually either the Fore or Back Coal of Kellie.

In the east of Fife, the most continuous exposures of rocks of the Lower Limestone Formation are seen on the shoreline between Pathhead and St Monans.

At the base of the Lower Limestone Formation, the St Monance Brecciated Limestone is seen on the shoreline below the nursery at Pathhead. This bed is widely recognized in the region from its fauna and characteristic broken appearance. Between the St Monance Brecciated Limestone and the topmost Upper Kinniny Limestone there are three other recognizable limestones: the St Monance Little Limestone; the Charlestown Main Limestone and the Mid Kinniny Limestone, all of which occur in separate upward-coarsening sequences with the intervening Mill Hill and Seafield Marine Bands. Sometimes a thin Lower Kinniny Limestone is recognized within the Lower Kinniny Marine Band.

As many as eight characteristic coals are identified: the four Radernie coals near the base; the two-leaved *Largoward Black Coal*; the divided *Largoward Splint* and *Parrot Coals*; the *Main Coal*; and the *Mid Hosie* and *Main Hosie Coal* below a ninth seam, the *Marl Coal,* which rests upon a marine mudstone that overlies or locally replaces the discontinuous Upper Kinniny Limestone. The boundary between this limestone and the marine mudstone above it is defined as the top of the Lower Limestone Formation (Forsyth and Chisholm, 1968). An attempt to summarize what is understood of the relationships between the various stratigraphic units incorporating details previously published by

Landale, Geikie, Forsyth and Chisholm, and Kassi *et al.* is given in Table 3. The close depositional relationship between the coals and the overlying limestones or marine bands may also be seen in this summary diagram.

In their accounts of the area Landale and Geikie were principally interested in the distribution of the coal seams, whereas Forsyth and Chisholm (1977), were more concerned with establishing the stratigraphic relationships by means of the limestones and marine bands. In the latter account they discussed the identities of possible equivalent horizons in the west of Fife. Importantly, they drew attention to the fact that the succession of sediments in the East Neuk varies in thickness from no more than 125m at the shoreline, rising to an estimated thickness of between 225m and 250m in boreholes at Lathallan and Radernie, 8km inland, before thinning to about 150m in the Drumcarro coalfield 4km to the north.

The most striking feature of the Lower Limestone Formation outcrops as a whole is the fact that they consist of a number of often closely folded and faulted, almost disconnected, basins, scattered from Pittenweem and St Monans in the southeast, west to Earlsferry and thereafter northwards past Colinsburgh and on towards the Lathones Fault at South Baldutho. The coal-bearing sequence continues north-westward through Balcarres, Largoward and Falfield before turning westwards to New Gilston and Teasses. Thereafter the coals may be traced north to Craighall, Gathercauld and Ceres. The final group of coal workings, which are limited to the east by the Radernie Fault, extend south-eastwards from Greigston through Peat Inn to Radernie and Brewster Wells.

The Pittenweem–St Monans Coalfield

Currently two spellings are in use for the name of St Monans village. To avoid confusion, the name used here for the village is as used by the local population and also the Ordnance Survey, in recognition of a local Christian leader, referred to as a saint, who established the settlement and the church beside the sea. The story of Monan is far from clear (Knight, 1933, in Towill, 1983). He is thought to have been a Dominican missionary from Ireland with Moinend, the companion of Brendan of Clonfert. He is reputed to have been killed, beside many Fifers, during a 'Danish' raid in AD875, allegedly the same raid during which St Adrian was slaughtered in the offshore settlement on the Isle of May. The 'Danes' here came not from Denmark but from the part of England known as the Danelaw (Dr Barbara Crawford, pers. comm.). The original church was largely destroyed at that time but a fine surviving medieval stone-built replacement was constructed on the original location in the twelfth century during the reign of King David II.

The name 'St Monance' as used by Landale and the British Geological Survey is perpetuated in an anglicized version of the proper name. This version is applied

Table 3: The Sequence and Thicknesses of the Coals and Intervening Sediments in the Pittenweem–St Monans Coalfield.

Name	Thickness	Description	Comments	Strata/Depth
Two Foot Coal	0.5m	Uppermost coal is a splint		Strata 7.0m
My Lord's Coal	1.5–1.8m	In 3 leaves. Most probably unworked	Lower leaf: poor quality. Top: good quality household coal. 7s6d–9s per ton	Strata 20.1m
Foulhouse Coal	1.2m	2 layers: Upper; splint; Lower: cherry	Cherry strong-burning clean coal. 7s6d–9s per ton	Strata 17.2m
Brassy Coal	0.9m	Good quality cherry; blacksmiths favourite	Only worked near Tofthill Pit. Much remains in ground	
Parrot Coal	2.1–2.7m	Upper part: parrot. Rest cherry or splint. All good quality. Exploratory drive 300m along level in seam before closure.	Travels well. From pit N. of Colinsburgh Rd. between Greendykes & North Coalfarm. Only coal worked on N.W. side of basin.	Grey, White & Blue stone 1.5m
Back & Fore Seams	1.2m (Back seam)	Back: ½splint ½cherry Fore: splint Seams separated by 2.0m of blue shale	Both cherry and splint coals in demand for export	Strata 17.1m
Mid Coal	0.9m	Soft cherry coal	Sold well for household use. 7s6d–9s per ton	Strata splint 11m
Wanderer Coal	Not plotted	No description		Strata: 3m
Four Coals: 1st seam	1.2-1.8m	2 leaves: splint; cherry Worked together	Splint: good for export; cherry: blacksmiths' preferred coal.	Blue stone 0.6m
Four Coals. 2nd seam	1.8m	2 high-quality seams worked together	On S-E side of basin near Standing Stone	Blue stone 1.2m
Four Coals 3rd seam	0.9m	Good quality splint	Worked at same pits as 1st & 2nd seams	Blue stone 1.8m
Four Coals 4th seam	0.8m thick	Hard splint	Little worked mainly near Tofthill Pit	Strata sandstone 11m
Four Coals 5th seam (Thirteenth Coal)	0.9m thick	Cherry coal: good burning household coal	Mainly worked when windmill drained pit, less when drained by steam pump.	Strata 10.1m
Four Feet Coal	1.0m thick	Household coal	Substantially below overlying sequence	

to define specific geological features such as the St Monance Brecciated Limestone and the St Monance White Limestone.

Although there are traditions of the medieval coal workings in eastern Scotland being undertaken by the Cistercian and Tironensian monks of the various ecclesiastical establishments, there is no known record of the monks from the Benedictine Priory of Pittenweem having exploited the coals to the west of the town. An early outpost of the monks was their navigation beacon on the Isle of May, which was supplied with fuel from an unspecified priory on the mainland until 1318, when it came under the control of the Augustinian order based at the St Andrews Cathedral Priory, who closed it down. Later the navigation beacon and lighthouse were fuelled by coals from Wemyss.

The exploitation of the coals and the use of salt pans, often in combination, are known to have occurred along the Pittenweem shoreline during the Middle Ages (Whatley, 1984, 1987) and at least the salt pan workings were controlled by the monks from the local priory (Paula Martin, 1999). Several sites were feued for the pans in 1530 and for the coals in 1535. Three feus on the grounds of the Pittenweem Priory are known to have been issued in the late 1550s to Andrew Wood, Hugh Moncrieff and Andrew Simpson for exploitation of the minerals, i.e. coals (Sanderson, 1982). These documents, accessible in the Special Collections of the University Library in St Andrews, show that Balfour, Commendator of the Priory of Pittenweem, feued the coals from the Acredale Lands area west of Pittenweem village in 1567. The coals were not to be transported elsewhere amid fears that coal was becoming scarce. At least two salt pans were in use in 1594. The coals of the basin were later worked as part of the Kellie estate by the Earl of Kellie, who believed the collieries had been in operation for at least 300 years by that time.

In 1655, when the estate was seized under the orders of Cromwell the mines were principally level-free. It became necessary to use horse-driven pumps to remove the surface waters to drain the workings below the level of high tide on the adjacent coastline. It was said that the working using a horse gin continued for some years under the supervision of Thomas Tucker. The lack of adequate pumping ensured that some of the coal workings were abandoned. The original but undated coloured 1795 map produced by Hogg, Estate Factor, for the Interlocutor indicates a dispute between Sir John Anstruther and Sir Ralph Anstruther (Fig. 10.9) shows evidence of the former existence of at least twelve bell pits to the southeast of the horse-driven gin pit south of the A917 road. Crop marks recorded by Colin Martin (1999) in aerial photographs have enabled identification of at least five sites used for horse-powered pumps in the coalfield.

Once the Kellie estate regained control of its lands, the exploitation of the coals was resumed largely to the north of the present B942 road towards

Abercrombie. By the 1730s and 1740s the Earl of Kellie had removed as much coal as was deemed possible by means of day-level drainage, and according to the First Statistical Account working of the coals was abandoned for some years. The Earl approached Pittenweem Town Council, gaining permission to divert waters from the Dreel Burn to power a water wheel to operate a pump for removal of the waters from his coals. The level of success of this pump was not great and these workings were abandoned for some years. The routes of the diverted stream are shown on Hogg's plots of the coalfield in 1795 (see Fig. 10.9).

The Newcomen 'atmospheric' engine, which was developed in 1707 to assist in draining the tin mines in Cornwall, did not arrive in Fife until over half a century later. The first steam-operated pump was installed in the Pittenweem coalfield in 1764 at Coal Farm, but this pump was not a total success, its capacity being less than was required, and the workings were later improved by the Newark Salt Company in 1775.

In 1753 Sir John Anstruther at the age of 38 inherited the estate of Elie and became the second Baronet. He proceeded to purchase many adjacent properties including the Kellie estate in 1770 and the limestone quarries along the shoreline between Pittenweem and St Monans. These properties gave him the feudal superiority of Pittenweem and a seat on the town council. He further strengthened his position when he ensured that his factor, Gavin Hogg, whose brother-in-law, Robert Fall and others from his estate were all appointed to significant posts with the council. Hogg's map of 1795 shows that Sir Ralph Anstruther of Balcaskie, (also a second Baronet), purchased at least nine of the adjacent rigs of land to be used for agricultural purposes.

At this point it may be important to point out that there were two baronetcies attached to the name of Anstruther. First, Sir John Anstruther, Baronet of Elie (1678–1754) and his eldest son, also Sir John Anstruther, second Baronet of Elie (1718–1799). The second family with whom the first family did not always see eye to eye, was that of Sir Ralph Anstruther, second Baronet of Balcaskie, followed by Sir Robert Anstruther, third Baronet (1733–1818), whose eldest son, also Robert, a Brigadier-General, died at Corunna in 1809. The fourth Baronet was Sir Ralph Abercrombie Anstruther (1804–1863), the fifth, Sir Robert Anstruther (1834–1886) and the sixth was Sir Ralph William Anstruther (1858–1934). Both families observed the long-standing Scottish tradition of passing Christian names from generation to generation. This can lead to confusion when comparing dates and names involved in industrial development and disputes.

In 1771, when the demand for salt was high, Sir John Anstruther and Robert Fall set up the Newark Coal and Salt Company and in the same year petitioned the council for permission to repair and enlarge Pittenweem Harbour. In return, they were to be charged no dues on shipping carrying their wares into or out of the port and to have free anchorages and preferential berthing in the new part of the harbour. The earlier storm-damaged harbour walls were duly replaced and alterations completed.

Nine new salt pans were constructed at St Philips between 1772 and 1774 (Paula Martin, 1999), and in 1773 the Newark Company sought permission to construct a timber wagon way from the colliery and saltings to the harbour at Pittenweem, crossing some of the town-owned land on the way. The numbers and sizes of ships using the harbour increased substantially. The larger vessels that were associated with the newly developed trade with the Low Countries, sailing to Amsterdam, Middelburg and Ostend, became regular callers when French salt was no longer available to them as a result of international conflicts. The Second Statistical Account of Pittenweem indicates that the salt pans were no longer in use in 1845.

Leitch and Neil wrote that they visited the colliery while it was still active and discussed its workings with Gavin Hogg, who was firm in his belief that the mines had been worked 'for generations before' the end of the eighteenth century. Paula Martin (1999) showed that after the colliery was leased to the Newark Coal and Salt Company in 1771, coal production was over 2500 tonnes in 1772 with 400 tons going to export. In the peak year of 1778 over 4300 tonnes of coal (1500 tons for export) were produced by a staff of over 300. There was a sharp fall in the annual exports to around 500 tons between 1779 and 1783. The total coal production fell from 3600 tons to 2600 tons in that period, revealing the importance of the local markets. Early in the 1780s, the original Newcomen steam driven pump installed in 1764 had been replaced by a Watt double-acting rotary steam engine after its predecessor had been working well below design capacity for some years. In a rapid upsurge in activity, the mine was deepened and coal production recovered to reach a new peak of 3700 tons in 1787, with annual exports reaching over 2600 tons in the period 1787 to 1789. Thereafter there was a steady decline, with the export trade virtually ceasing by 1795 following a major fire below ground (Paula Martin, 1999).

In 1794 Sir John indicated to the council that he would be giving up using the wagon way and reducing his maritime commitments (i.e. reducing his export of coals) after one more year. After that time elapsed, the wagon way ceased to be used, but for some years after that the council maintained the facility in case the coal industry became re-established.

There is little to see of the wagon ways themselves today, but under suitable conditions, crop marks indicate their routes across fields, as marked on Hogg's map (Fig. 10.9) and displayed in air photographs taken by Colin Martin (1999) while investigating the salt industry related structures at the St Philips saltings.

He has most kindly granted permission for the author to use two examples in Figures 10.11 and 10.12.

Colin Martin has identified traces of the two former timber tracks to Pittenweem harbour and pointed out the remains of many early industrial buildings and sites of winding engines in the area (Figure 10.11). It is always difficult to be assured of identifications of structures from such material, but the crop marks suggest industrial development from south of the A914 road from Tofthill Farm to Coal Farm and north towards Waterless Bridge, whence along the west–east line of the wagon way 3.5km north of the road junction at Croft Hill (N56.2130, W2.7449). The scale of the workings in this coalfield may be gauged from Galloway's map (1895) (Fig.10.10), and the fact that more than five clear sets of winding wheel (horse gin) foundations have been readily identified, albeit in various states of decay (Figure 10.12). Some of the former mine workings and geological features are readily identified on each of the plates.

But what of the coals themselves? In this coalfield we are no longer dealing with the two major and two minor seams as at Kellie, but with more than a dozen

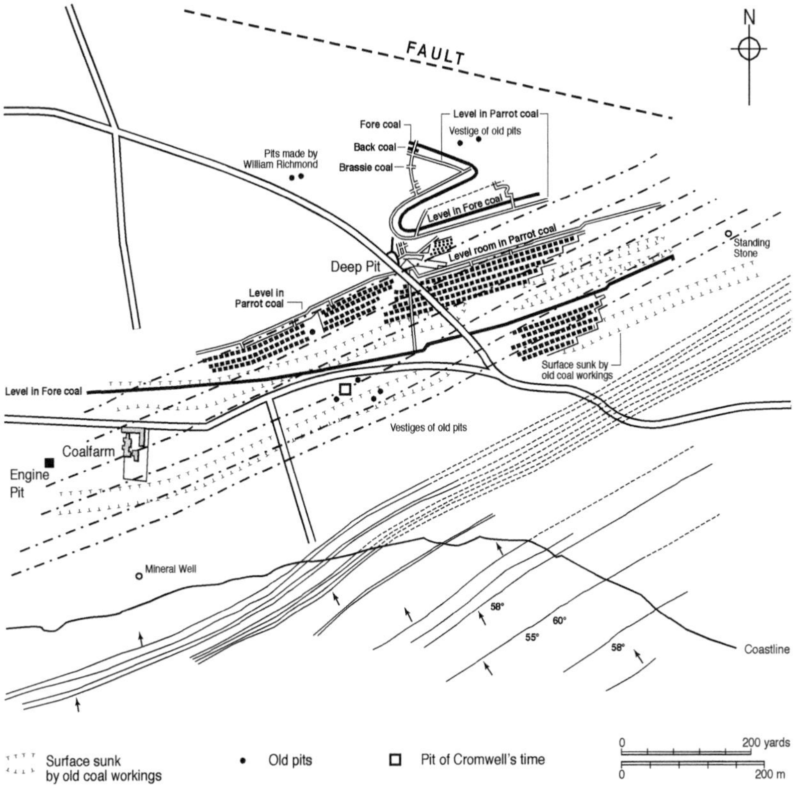

Figure 10.10 Galloway's map of the workings of the Pittenweem Coalfield, 1895 (adapted by Graeme Sandeman).

Plate 10.11 Aerial view of crop marks at Pittenweem showing walking circle of the horse and outlines of building foundations (photograph Colin Martin).

Plate 10.12 Aerial panoramic view of the Pittenweem Coalfield (photograph Colin Martin).

individual coals, albeit only ten of those were potentially workable. Following the provision of plans of the coalfield by Hogg in 1795, the first description of the coals was given by Williamson (1839), who named them in sequence and commented on their characteristics. His listings, largely followed in all subsequent reports, are summarized in Table 3 (see p. 95), starting with the *Four Foot Coal* at the lowest level and working up the succession to the *Two Foot Coal*. Landale (1854) provided detail of the names and actual coal thicknesses and also gave a measurement of the rocks separating the seams. The lists were enhanced by Geikie (1902) and Dron (1902) but later modified by Forsyth and Chisholm (1977). Confident correlation between the two tables does not appear possible today. Unusually, none of the early names appear to have survived. The listing as notified by Geikie (1902) is given in Table 3 (see p. 95), which also shows the thicknesses between the coals.

How are the coal seams arranged in this basin? The syncline enclosing the whole of the Lower Limestone Formation, overlain by a core of the lower part of the Limestone Coal Formation, is almost constantly 1km in width from northwest to southeast. The fold axis may be traced for around 2.5km plunging gently towards the north-east. It is terminated in the north by the northeast to southwest-trending major Ardross Fault, which is believed to be present close to Waterless Bridge and continues north-eastwards a short way north of the B942 road. The overall structure is recognized as consisting of a double trough separated by a narrow central anticline. The moderate inclination of the north-westward dipping eastern limb decreases towards the main fold axis where the dips are reversed to form a sharp syncline. This is followed by the small anticline to the west, of which a second small syncline is recognized towards St Monans. Both Galloway and Hogg believed that the overall structure took the form of an open structure with moderately inclined limbs (Fig. 10.13A). Hogg gave a rather figurative, linear interpretation of the structure, which he believed to be largely symmetrical. However Geikie (1902) following Leitch and Neil, provided cross-sections of the syncline correctly, showing the presence of a steeply dipping western limb that would have been of nearly vertical edge coals on that side of the basin, in contrast to the moderate dips of the eastern side (Fig. 10.13B). Landale (1839) believed this

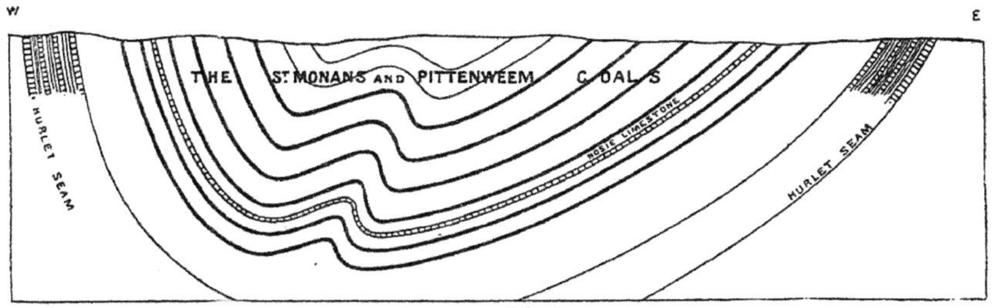

Figure 10.13A Cross-section of the Pittenweem Coalfield according to Hogg.

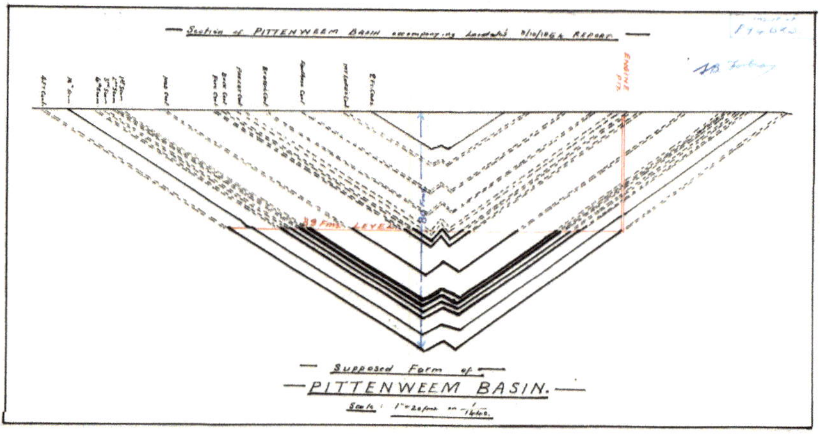

Figure 10.13B Cross-section of the Pittenweem Coalfield according to Geikie.

lowermost coal to be from a much lower level than the lowest exploited coal listed above, and pointed to the presence of two thin limestones, which he suggested might equate with St Monance White and the Upper Ardross Limestones with their attendant thin coals within the top of Pathhead Formation, but there was no sign of the otherwise ubiquitous St Monance Brecciated Limestone nearby. In total, the coal-bearing succession of the Lower Limestone Formation in this area is believed to be approximately 120m in thickness.

Most of the early mining was undertaken on the eastern Pittenweem side of the coalfield and consisted initially of working downwards from the land surface, especially in the Tofthill Farm area. The extraction became extended into shallow underground room and stoop workings, which often left the land surface heavily disturbed as behind the advancing excavations, roof collapse of the abandoned rooms led to the ground surface becoming broken so that, until fully reinstated, it was unsuitable for any agricultural activity. In response, in the late seventeenth century and early eighteenth century, the Pittenweem Sea Box Society reduced the land tenancy rents to those whose land had suffered damage.

Two zones of depression of the land surface were marked on Hogg's plan of the coalfield (Fig. 10.13A).and reported by Galloway (1882). From a knowledge of the succession of coals as described by Williamson (1839), all of which dip north-westwards from the eastern margin of the coalfield basin, it will be seen that the early noted depression along and close to the modern A914 road between Pittenweem and St Monans is most likely to have been related to near-surface workings on the First, Second and Third Seams of Four Coals. A second depression a short distance to the northwest probably marks the location of the surface and near-surface extraction of the Foulhouse, Parrot and the Back and Fore Coals.

According to Landale (1854) the early fittings of the mine carried workings down to a depth of 11 fathoms (20m) from the surface. This enabled the extraction of the coals above the level of high water, allowing the mines to be drained by day levels discharging to the coast. Some of the earliest drainage was by water-wheels driven by waters diverted from the Dreel Burn, as arranged by the Earl of Kellie. Later several watermills were installed to help remove the waters, which were discharged on the surface, draining towards Pittenweem. It was not until 1764 that a Newcomen 'atmospheric' pump engine was installed enabling the full lower depth of workings to be brought into use. A second set of fittings carried the workings down a further 27 fathoms (49.4m), to a total depth of 38–39 fathoms (approximately 70m), and all profitable seams were taken down to the level of this Deep Pit. The waters from this lower level were lifted by the pumps in the Bye Pit at Coal Farm (N56.2090, W2.7543) and discharged into a pre-existing higher day level. A line of pits marked on the original Hogg 1795 map (not the one used in subsequent legal proceedings), shows them to have been 10m west of the track to Pathhead (N56.2108, W2.7369) on the line that includes the 'mineral well' marked on the Ordnance Survey maps. Although the pump continued removing the waters until final closure of the mines in 1803, its efficiency fell steadily and it required increasing maintenance to enable the levels to remain at their design depth. The waters from the 'mineral well' might have been believed to have restorative properties by travelling gullible folk prone to visiting spas elsewhere, however this particular well is reputed to have been used by the local fishermen for curing their nets in addition to carrying waste waters from the mines.

Galloway (1882) indicated that some westward exploration had been achieved by driving a stone mine through unproductive rocks across the structure at the 70m level, but it was not clear when or by whom this was undertaken as it does not appear on the original Hogg map. The stone mine appeared to confirm the continuity of the syncline but failed to penetrate sufficiently into the eastward-dipping western limb of the fold to demonstrate which, if any, of the coals were present and workable on the western side of the basin. The Hogg map also marks the position of two pairs of pits put down at an unknown date by Richardson about 280m northwest of the Deep Pit but before the Pittenweem workings became fully established. There is no indication of any coals having been extracted from these pits, but a further trial pit 120m to the east indicated that it entered mine wastes before being abandoned, so there must also have been some earlier workings in that part of the basin.

In his initial account of the coal basin, Landale (1854) indicated that in addition to recognizing the Harbour Coal, he had searched along the steeply dipping western margin of the syncline for any of the other coals seen elsewhere in the basin, but located only two. These he and Dron (1902) identified tentatively as accompanying the White and Abercrombie Limestones seen elsewhere towards the centre of the basin. The BGS staff mark two limestones on their 1:50,000 scale published map identifying them as the St Monance Brecciated (Hurlet) and Charlestown Main (Blackhall) Limestones. No further reference to the coal exposures has been located. In the absence of firm proof in the form of fossil evidence, it should not be forgotten that, as both Landale and Dron pointed out, within the Pathhead Formation similar limestones with associated thin coals are exposed on the beach east of Pathhead.

The very substantial Ardross Fault extends from Elie Harbour in the west and follows a path to the north of St Monans, passing Waterless Bridge and Anstruther and on to Kilrenny. It truncates the western limb of the syncline. Any steeply inclined coal seams close to the line of fracture could have deformed by shearing and effectively lubricated the lateral movements of the adjacent rock masses. In the absence of exploratory cross-cutting mine drives or a well-designed pattern of boreholes to establish whether the coals are actually present about this north-western side of the syncline, there must be serious doubts on the convictions of many local people that workable coals are still present.

The closure of the coal mines at Pittenweem led to great public distress as the pits, when fully operational, were believed to have provided work for over 300 people in an area in which non-maritime employment opportunities were scarce. Public meetings were held to protest the closure and to urge that work on the coals should restart. Despite considerable good will to search for funding to continue the work, no progress was made. Nearly 50 years later the idea of restarting the mines resurfaced, as reported in the Pittenweem Register, but to no avail, and it was another 40 years before the matter was again discussed in detail. These later meetings were fully reported in various local newspapers and as a result, Robert Galloway, Professor of Geology from the University of Cardiff, was appointed to review the geological knowledge of the site and make recommendations. His fully documented and detailed report, including estimates of tonnages of coals that might remain in place, together with calculations of the volumes of water likely to require to be pumped out, was duly published in its entirety in the local press of 26 August 1895. In conclusion, he urged caution and stressed the need for a programme of exploratory boreholes to assess the likely success or otherwise of the enterprise. The report was well received, but no further progress or sustained programme of exploration for coals towards St Monans has since been undertaken.

The core of the syncline between St Monans and Pittenweem shown on BGS maps notes the presence of the Limestone Coal Formation, but there is no indication of any coals having been recognized outcropping at the land surface within this area.

Pathhead to Earlsferry

As can be seen on the foreshore to the west of Pathhead, the rock succession is rather more varied in the Lower Limestone Formation than in the preceding Pathhead and Sandy Craig Formations. The repeated sequences of the sediments, reflecting systematic variations in response to changes in the ancient sea levels, have been examined by many workers following the original detailed records of Kirkby (in Geikie, 1902), with later contributions from Tait and Wright (1923), Goodlet (1957), MacGregor (1968 et seq.), Fielding *et al.* (1988), Francis (1991), Hooper *et al.* (2002), Underhill *et al.* (2000) and Kassi *et al.* (2004). The more recent works have inevitably built on the findings and interpretations of the earlier authors according to the beliefs of the day.

In order to assist in the understanding of the rocks exposed on the shoreline between Pathhead and St Monans harbour, it may be helpful to refer to a summary of diagrammatic sediment logs from Kassi *et al.* (2004) as presented in Figure 9.2. The two columns indicate the thickness and lithology of the different rock types from the 4.8m-thick St Monance White Limestone, the last limestone in the Pathhead Formation (at the base of the left column) upwards past the *Hurlet Coal* to the base of the St Monance Brecciated (Hurlet) Limestone. The base of the thin intervening claystone layer is the start of the first major flooding event by the sea at the start of deposition of the Lower Limestone Formation. The sequence continues upwards along the right-hand column to the Upper Kinniny (Top Hosie) Marine Band, which marks the top of the discontinuous Marl Coal at the base of the overlying Limestone Coal Formation.

In each, the left column indicates the rock type and height above the St Monance Brecciated Limestone, and to the right, the breadth indicates the texture, with narrow mudrocks, clays and shales, broadening gradually as the rock material becomes coarser, to thick sands and limestones. Various internal structures such as current bedding, burrows or fossils present are also indicated.

A first inspection of Figure 10.10 reveals that there are no fewer than 15 coals present, most of them as thin seams. The coals are grouped into two parts of the succession. Near the base are the rocks typical of the Radernie Coals and in the upper half are the Largoward Coals. The coals vary in texture, composition and quality throughout the coalfields concerned, and their characteristics will be discussed at each locality rather than solely in relation to the succession exposed at the shoreline. Limestones and mudstones of the Marine Bands are also recognized throughout the succession; the former having been worked for agricultural lime in most of the coalfields.

To the west of St Monans Harbour the exposed rocks along the 4km of the foreshore as far as Elie Ness reveal small isolated masses of the Pathhead Formation rocks, such as the Ardross Limestone, identified by their fossils.

In addition, there are at least six separate basaltic ash and agglomerate-filled volcanic necks with basalt dykes on the south side of the Ardross Fault line (Francis and Hopgood, 1970). The volcanic ashes and agglomerates consist of darkly coloured, irregularly shaped fragments of broken lavas mixed with occasional locally derived pieces of sedimentary rock, all of which may vary in size from dust to pebbles and boulders. In places the fills of the former volcanic necks show steeply inclined internal layering developed parallel to the margins of the neck in response to the streaming of escaping gases. The structure of these volcanic necks was also addressed in some detail by Geikie (1902) and, although of considerable scientific interest and well worth investigating, the detail will not be repeated here.

The age of the volcanic necks is not firmly known, but it is believed that they were injected along the deep-seated fracture of the Ardross Fault system, the southern side of which is displaced towards the southwest. Some vertical movement is also thought to have taken place between periods of vulcanicity. The displacements along the faults are thought to have taken place very late in the Carboniferous, or possibly just into very early Permian times around 300 million years ago.

The prominent headlands of Sauchar Point and Elie Ness are prime examples of features formed on the volcanic necks. On the margins of Woodhaven Bay, the wave-washed surfaces are normally well exposed. ***Do not attempt to kick apparently loose pebbles on the surface of the wave-cut platforms: many of them are still firmly attached to the bed rock.*** Ruby Bay, a local name for the area between the two headlands, is the site from which dark red garnets, locally known as Elie Rubies, have been found in the past set in the volcanic materials exposed during low water of spring tides. The minerals, although attractive, are not genuine rubies, having a different crystalline structure and chemical composition. The garnet family are all of complex silicates forming crystals of cubic form, whereas the genuine rubies are of the corundum family of minerals, all of which are formed of aluminium oxide and have rhombohedral crystal form. Corundum is a natural mineral that is second to diamond in its hardness and is often used as an industrial abrasive.

Prominent exposures of the basal St Monance Brecciated Limestone and the overlying Charlestown Limestone are encountered before Elie Harbour is reached.

Earlsferry and Grange

The Earlsferry coals are exposed on the foreshore below a mobile beach sand cover at the western end of Elie Bay beside Capel Ness. The thinly bedded sandstones, siltstones and both the Mid and Upper Kinniny Limestones dip westwards at a low angle. Amid the extensively current-bedded sediments are many plant fragments of twigs, leaves, broken pieces of bark and trunks of several species. ***Please observe without destroying them.***

The mine records show that the northward track across the golf course towards Grange Farm (N56.1916, W2.8261) leads over at least three east–west trending faults, each of which has a northward downthrow. The vertical displacement on the southern fault near high-water mark is not known (Landale, 1837). The mid-links fault moved the rocks down for 59m (Geikie, 1902, p.164) and Dott (1944) believed that the movement on the third fault, north of Grange Farm, reached 55m. North of this fault the rocks are heavily fractured, limiting continued exploitation.

Where thin igneous intrusions travelled upwards along the fault planes, the marginal coal seams were turned upwards to give vertical and near-vertical dips beside the intrusions. When these edge coals were bored into from the surface, they suggested great thickness for the coals. Alas, when trial pits were cut, the coals were found to be no more than 2.4m thick, the boring having penetrated along the length of the upturned seam rather than cutting directly across the breadth of it.

After the faulting an intrusion of basaltic rock cut across the rocks of the Limestone Coal Group to form the headland of Chapel Ness. A complex of variable volcanic ashes, agglomerates and basalts continues along the cliff to Kincraig Point and beyond Shell Bay to Ruddin Point.

The workings in this coalfield are said to have been in operation as early as the sixteenth century (these probably near the shoreline), and are known to have been successfully continued into the 1850s. Mr Keddie (Junior), partner in a legal firm in Cupar, who took on the lease of a coal working at Grange in the mid-1820s (H. Keddie, 1922) appointed his very experienced collier father (Keddie, Senior), to oversee the workings at Grange. The recognition that the mining activities had caused considerable disruption to the land surface through sinking of pits, erection of machinery, creation of coal hills, etc., led Keddie (Junior) to extend his lease in 1843 to include Grange Farm itself. Father Keddie instituted a programme of exploration boring to locate further coal reserves. No substantial new seams were located, but a layer of ironstone was recognized as potentially as valuable as the deposits being worked about that time at Winthank (McManus, 2010) although the irregular distribution of the ore rendered it not worth pursuing.

The workforce was faithful to the Grange workings, supporting their overseer in a court case, but the management faced an unexpected problem when much of the workforce deserted the mine workings during the summer months in the 1830s and 1840s to follow the seasonal North Sea herring trade (Michael

Martin, *et al.*, 2007). This was of considerable importance, as the production of coals during the summer months normally permitted the creation of coal hills to supply heavy demand for domestic coals during the winter months.

Landale (1837) considered that there were no fewer than 17 coal seams in the area, a figure suspiciously similar to that known at Pittenweem, (Grant Wilson of the BGS in Geikie, 1902). Andrew Rolland, believed by Landale to be one of the more reliable former miners, provided a useful succession of nine coals in the Grange area (Table 4).

Table 4: Coals in the Grange workings after Rolland in Landale (1837).

Name of Seam	Strata	Thickness
Top Coal	5.5m	3.35m
Under Coal	Uncertain	1.2m
Three-feet Coal	"	0.9m
Cherry Coal	"	0.9m
Cherry & Splint Coal	"	1.5m
Two-feet Coal	"	0.6m
Two-feet Coal: Blaes	"	0.36m
Two-feet Coal: Ironstone	"	0.41m
Thin Coal	"	Uncertain
Main Coal	47.5m	1.5m
Salt Coal	Uncertain	0.9m

The *Main Coal* was estimated as being 9.1m below the *Cherry* and *Splint Coal.* The 0.9m thick *Salt Coal* was 3.7m above the Johnstone Shell Bed (Francis in Forsyth and Chisholm, 1977) and Forsyth and Chisholm (1968) linked this coal to a 0.97m thick seam at 207.4m down the Dunotter Borehole (N56.2120, W2.8641) near Muircambus, some 3km to the northwest. The latter authors were doubtful about the quoted thickness of 47.5m between the Main and Salt Coals.

On their map of the coalfield, Michael Martin and his colleagues show it to have supported three pits and a housing for a steam engine to the east of a footpath south from Grange Farm and Grangehill to the west end of the village. To the west of this track they showed one further pit, but no details of the actual workings are known to exist. According to Geikie (1902) the pits on the links were variously 36.6m and 54.9m down to the Main Coal and another, presumably to the south of one of the main faults, penetrated 47.5m to the Salt Coal. The principal coals worked were the *Main* and the *Cherry* and *Splint,* each of which was taken from rooms off two separate levels. Both were of consistently good quality when followed north-westwards along the strike direction. When traced downwards in the dip direction they quickly degenerated, as they had been increasingly subject to the heating and disruptions associated with the nearby igneous activity at Kincraig Point. Elsewhere the coals maintained their quality as good household coals, which sold well in the neighbourhood.

The Output Book of the Grange Colliery 1826–1830 (L622.33) in the reference section of the Cupar Library reveals the names of the workforce, the numbers of days that they worked in the colliery, the production achieved from the different coals, and details of the daily accounts of the Grange Colliery. Once the pattern of working had been established the 5–10 colliers were working 10–12 shifts each fortnight, and this produced 125.6 tons (113 tonnes) of saleable coals with a further 256 tons (234 tonnes) of unsold coal on the hill. A year later the production had increased to 270.7 tons (245 tonnes) with a further 1527 tons (1335 tonnes) stored on the hill. An additional 44.9 tons (40 tonnes) of coal had been used in two weeks to fuel the new engine. The workforce in the autumn of 1836 consisted of 19 colliers and 13 labourers, but these numbers changed as demand for the coals increased during the winter and spring months (December to April) so that in the first three months of 1827, 22 colliers were listed as being employed. Of these 5 worked only one shift but they were supported in dealing with various tasks around the colliery by 18 labourers. The wages earned by the colliers varied according to the conditions faced below ground, but the men working for 12 shifts in a fortnight would normally receive £4 2s 7½d.

In the winter and spring the demand for coals for domestic use rose sharply and, as running costs remained relatively stable, so the fortnightly difference between sales and expenses were healthy at that time of year. In the summer months of May through August, when the sales were lower, the profits fell, sometimes to below £6 in each two-week period of counting. The sales price of the Splint Coal varied between 5s 0d and 6s 3d per ton while the less popular Cherry Coal sold for 3s 9d to 5s per ton during the three years 1826–1828.

There is no indication that the numbers employed at the colliery ever approached the 90 indicated in the First Statistical Account as working in the industry within the parish of Kilconquhar, quoted by Michael Martin and his colleagues. Presumably there were further colliery workings elsewhere within the parish, possibly at Pitcorthie or Balcarres.

Beyond grouping the wages bill for the colliers into one large number at the start of the relevant pages, the accounts show that several of the colliers were paid for on-cost services such as redding (cleaning) levels and rooms, removing snow from colliery roads for domestic customers (March 1827), operating the windlass, tending horses for the gin pits, and redding the engine sump. It is good to note that the carters were not forgotten and allowance was made for their whisky on New Year's Day 1830. These same carters were required to pay tolls for their use of highways; the tolls for six weeks cost £4 9s 7d in September 1826.

The Grange Colliery was generally believed to have been a relatively safe operation as the Main Coal was said to have had a strong roof, but it is reported that Andrew Herd, who was serving as a hewer, was killed outright in 1851 when a large mass of roof rock collapsed directly on him where he was working. He left a widow and six children.

Coalfields of the Largoward District

When he was introducing the geology of the Largoward district for the first time there was a salutary warning from Landale (1837, p.303): 'This is a most complicated and convulsed district and I almost despair of being able to give the reader even a general idea of it'. Today, following much subsequent endeavour by the coal mining industry, several interpretations by BGS officers and detailed analyses by consultants for the National Coal Board, the major features of the geology of the area are much better understood. In the following sections a further attempt will be made to summarize the structures and their influence on the development of the coalfields.

The Lower Limestone and Limestone Coal Formations deposited in the 4 million year period between 329 and 325Ma ago are the principal coal-bearing parts of the geological column in eastern Fife. According to Forsyth and Chisholm (1977) the aggregate thickness of the two Formations in this area is probably little more than 450m. In the Ceres district alone the Limestone Coal Formation is known to contain at least 22 coal seams, and 15 seams have been identified in the Dunotter borehole drilled 1km southwest of Colinsburgh (Forsyth and Chisholm, 1968). By contrast there may be as few as 7 seams in the Lower Limestone Formation.

In that both formations have together been folded, faulted and intruded by sills and dykes and pierced by literally scores of volcanic necks, it may simplify analysis of the mining problems faced throughout the western parts of the East Neuk if the two formations are considered together. Both of the stratigraphic units are seen together, and both are broken into narrow strips of land between a series of northwest to southeast-trending faults. Within each of these 1–2km-wide belts, the rock formations have been subjected to gentle folding to create adjoining shallow synclines and anticlines with mainly northeast to southwest-trending fold axes. In the east the folds have moderately inclined limbs with dips rarely exceeding 20°. The combinations of folding and faulting lead to frequent repetition of outcrops of the individual coal seams (the *Largoward Black, Largoward Splint*, the *Marl Coal* and the *Largoward Thick*), and limestones (the St Monance Brecciated, Charlestown Main and the Mid Kinniny) at the land surface. This rather simple double control has resulted in the development of many relatively shallow coal mines to exploit the same relatively few economically productive seams throughout the whole area. The deepest coal seam shown on the published cross-section of the BGS map is the Largoward Black Coal shown to reach down to 450m at the keel of the syncline in the Cassingray area, but this seam was of relatively poor quality and would not have been mined to that depth.

Although the great variety of structures are readily shown by the bright colours allocated to them on the BGS Sheet 41 map of the Lower Limestone and Limestone Coal Formations, both the folds and the faults are also present in the underlying Pathhead Formation, the Sandy Craig Formation, the Pittenweem Formation and older rocks of the Strathclyde Group. Although these have not been distinguished, these structures show their continuity on the map in the Lathockar–Gilmerton–Dunino area. However, there is no evidence of any physical dislocation between the upper and lower parts of the rock succession, and all have been subjected to the same history of tilting, folding, faulting and intrusion, albeit with less apparent intrusive igneous activity in the east.

Over the years of exploration and exploitation many of the individual coal seams, limestones and other marker horizons have been assigned different names in more than one locality. To add to the confusion, sometimes the identical name has been given to different seams in different localities. Initially these names were derived for individual mines or quarries but attempts have been made, largely by BGS geologists, to pull together the plethora of local names to help them to relate the successions to those of Central and West Fife. With no fewer than 29 names of the coals in the two formations and countless unnamed thin seams, some of which may thicken to become significant elsewhere, traps abound in attempting such correlations. See Table 5 below.

Structures of the broader Largoward area

Reference to the BGS 1:63,360 scale map sheet 41 of the area extending between Craighall Den (Ceres) in the northwest to Largo Law in the southwest and Dunino in the northeast and Arncroach in the southeast reveals a complex arrangement of the rock units. After the layered rock successions had been deposited, they, including their economically valuable limestones and coals, have been subjected to folding. Many of the seams have been cut short at faults, and in places, the igneous intrusions have baked, occasionally heavily mineralized, and locally completely destroyed the coal seams.

The map may be simplified by considering the areas between selected fault lines from the Lathockar and Radernie Faults in the east to the west–east trending Teasses, Branxton and Largo Law Faults in the southwest (Fig. 10.14). The outcrops of the northern segments of the eastern fault-bounded blocks curve towards the west; the southern fault lines curve to the southeast as they cross a late gently

Table 5: A simplified guide through the many names applied in the coalfields of the Leven, Largoward, Ceres and St Monans areas.

Leven & Central Fife	Largoward	Ceres	St Monans
Limestone Coal Formation			
Blairhall Main		Luncart: Mak-Him-Rich	Splint: My Lord's Coal
Cardenden Smithy		Ceres-2-Foot	
Seven Foot		Ceres Thick	Foulhouse
Lochgelly Splint		Ceres Upper 4 ft.	Black
Lochgelly Black Parrot		Ceres 6 ft.	Parrot
Black Metals	Thick Coal	Little Splint	Back & Fore
Glassee	Marl Coal	Bowanton	Mid Coal
Mynheer		Donaldson	Wanderer
5 Foot		North	1st of 4 Coals
		Ceres Little	
	Whitemire	Ceres 5ft. & L.4ft.	2nd of 4 Coals
Dunfermline Splint		Ceres Whin	3rd of 4 Coals
Dunfermline Under		Ceres Rum	4th of 4 Coals
Johnstone Shell Bed		Shell Bed	
Smithy		Smithy	13th Coal
Sulphur Coal			
Largoward Thick	Largoward Thick	Ceres Black	4ft Coal
Kinniny Limestone	*U. Kinniny Limestone*	*U. Kinniny Limestone*	
Marl Coal	Marl Coal	Ballfield Coal	
	Appleton Ell		
Lower Limestone Formation			
2nd Hosie Limestone	Mid Kinniny Limestone		
Mid Hosie Limestone	Lower Kinniny Limestone		
	Largoward Splint		
Mid Hosie Limestone	Seafield Marine Band		
	Largoward Black		
	Pilkin Coal		
Charlestown Main	Charlestown Main Lst.		
	Radernie Marl		
Teasses Main	Radernie Main		
Teasses Under Coal	Radernie Duffie/ Duffy		
Inchinnan Limestone	St Monance Little Lst.		
	Radernie Brassie		
Hurlet Limestone	St Monance Brecciated Limestone		

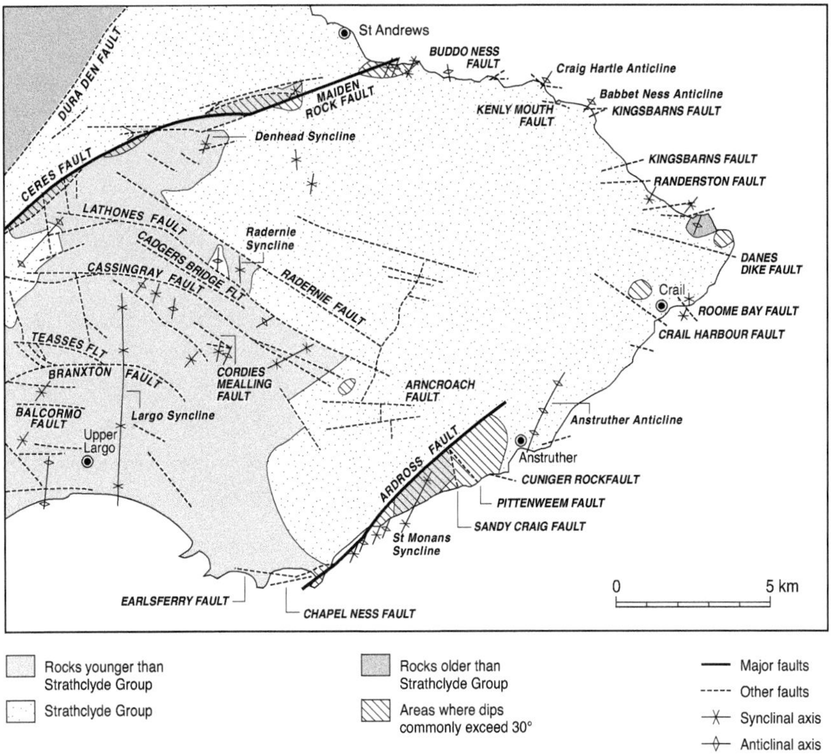

Figure 10.14 The main geological faults of the Largoward District (after British Geological Survey, 1977).

arched anticlinal fold axis. On each fault, the plane of displacement slopes steeply towards the southwest or south towards the downside of the fractured ground. The amount of vertical displacement on the faults is estimated by comparing the positions of two well-defined points of reference on opposing sides of the fracture plane. It may vary considerably along an individual fault. Along the Lathones Fault at Lawhead it is 91m, but 2km to the southeast at Lathones it is 270m. Likewise, on the Cadger's Bridge Fault, some horizontal displacement along the fault planes is suspected from the lateral displacement of adjacent fold axes.

The rocks exposed at the surface between the Lathockar and Radernie Faults are principally those of the moderately dipping Sandy Craig and Pathhead Formations. The only significant coals recovered from this area were the Fore Coal and Back Coal of the Kellie coalfield extending northwards from Carnbee.

Between the 10km-long outcrop of the Radernie Fault (between Kininmonth and Carnbee) and the 13km-long outcrop of the Lathones Fault (between Ceres and Wester Kellie), the rocks of the Pathhead Formation in the southeast form an anticline, the western limb of which dips in sequence beneath the base of the Lower Limestone Formation. Although the St Monance Brecciated Limestone and the early coals of the Lower Limestone Formation are recognized in the Radernie area, no indication of the presence of the *Largoward Splint Coal* appears on the map in this block until it is identified in the grounds of Greigston House (N56.2890, W2.8944). Traces of former workings of this coal may be seen as several widely spaced 2.5m–3m diameter round hollows of long abandoned shafts within roadside fields between Greigston and Blackwalls. There is a similar lack of continuity with the Charlestown Main Limestone present in the Radernie Mine area but not known on the western limb of the Radernie anticline. It is next marked north of Wilkieston 3km north of Radernie and east of the Radernie Fault. The rocks have been gently folded and small anticlines are present to the southeast and northwest of the Radernie syncline.

The almost horizontal Essexite sill north of North Baldutho is undisturbed as it crosses both faults and therefore post-dates movement on both. The microporphyritic olivine-dolerite to the southeast is also not displaced by movement on the Lathones Fault. The teschenitic dolerite at North Bowhill is not displaced by the relatively short Cadger's Bridge Fault but appears to be partly displaced by the Lathones Fault. To the north, several olivine dolerites and the Wester Radernie analcime dolerite are broken by both faults. This shows that there were at least two phases of intrusion, i.e. one before the last phase of faulting and the other after it.

The next fracture to the south and west is the Cassingray Fault, whose outcrop can be traced for at least 13km from Teasses to Wester Kellie. In the south the area between the two fault lines reveals the Sandy Craig and Pathhead Formations beneath the Lower Limestone and Limestone Coal Formations. A shallow, open south-westward plunging syncline and the adjacent South Cassingray anticline are formed principally within the Lower Limestone Formation. The two main formations alternate at the surface, marking a series of low amplitude folds between two bounding faults.

Beyond the bend of the fault line to an east–west direction lies a north–south trending syncline with the Limestone Coal Formation at its core, which passes down to the Pathhead Formation between Fleecefaulds and Craighall Den. The folded sedimentary rocks are terminated against the Lathones Fault to the north and the Cassingray Fault to the south. They and several olivine dolerite intrusions that are similarly terminated at the faults therefore predate both fractures. The quartz dolerite sheet that overlies the core of the syncline cuts across the northern olivine dolerite, which terminates against the Lathones Fault and is in turn broken through by a small volcanic neck.

Between the Cassingray and Cordies Mealling Faults, a similar pattern of anticlines and synclines with northeast-trending axes is maintained. However, south of the Cassingray Fault the south-easternmost anticlinal axis is displaced towards the northwest by 400m or so, but this displacement decreases towards the northwest. The thicknesses of the sediments forming the folds in this segment of the map are not great, for no units below the *Largoward Black Coal*, nor any of the seams above the *Largoward Thick Coal* are shown to be present, suggesting that a total thickness of no more than 120m is involved in the next surface exposures. Both bounding faults, which are closely associated with several small volcanic necks, are buried beneath the margins of a teschenitic dolerite sill.

Beyond the southern end of the Cordies Mealling Fault the olivine dolerite sill of Kilbrackmont Craigs is present across the line of the fault, and to the south of this mass the 2.5km Balmakin Fault follows a similar northwest to southeast line, breaking through the Lower Limestone Formation and into the underlying Pathhead Formation.

To the south of the Cordies Mealling Fault only one substantial fracture, the Gilston Fault, is present before the Branxton Fault is seen at the surface. This introduces the Upper Limestone Formation before reaching the Largo Law massif, where layered volcanic ashes and tuffs are interbedded with the limestones, shales and coals of the formation. The pronounced patterns of folding seen in the more northerly parts of the map are not as evident within this block, but a fine series of nearly concentric outcrops of the coal seams marks the presence of the Appleton basin between Baldastard (N56.2531, W2.9378) and Woodside (N56.2601, W2.9324). The Largoward Thick Coal outcrops in many places around Gilston House are the result of strong topographic variation. The frequency of volcanic necks, filled largely with tuffs and including small masses of various dolerites, including a Monchiquite on Largo Law itself, increases to the south. The evident volcanic necks, which are thought to be of Late Carboniferous age, break through all of the stratigraphic units but only some of the faults.

It is important to recognize that the absence of a particular bed or coal seam being marked on the geological map does not always signify that it is not present, rather that it has not been seen or identified beneath the sediment cover.

Chapter 11

The coalfield surveys by Dott and Forbes

During the Second World War when supplies of coal were vital in underpinning the industrial efforts of the country, various mining engineers were given the tasks of assessing the reserves of coal still in the ground. In the east of Fife, the task was given to George Dott, who had access to all previous reports on the development of the mines of the area. The opening comments of his report are both instructive and informative:

> The area is one of the most famous in the British Isles for the study of volcanic phenomena ... a notorious battleground for rival schools of geologists ... the Memoir of the ... Geological Survey of Scotland by Geikie suffers from too much vulcanism from the viewpoint of the Mining Engineer ... disappointingly little to say about structure ... no correlating the rocks ... all the coal detail comes from Landale's works ...

A copy of Dott's unpublished review dated 1944 is held in the Library of the BGS in Edinburgh. It contains many thoughtful assessments of the mining records with reference to the detailed Mine Abandonment plans held in the same library. Much of the information presented either provides the lengths of the surface outcrops or is in coded form relating to individual seams and their distribution within predetermined areas including coal types and thicknesses determined from boreholes. The level of detail needed as relevant to that Report is beyond the needs of this review. Following Nationalization of the coal industry in 1947, which transferred the responsibility for coal production in the post-war era to the National Coal Board, an extensive list of targets was identified whereby the Board could determine the reserves of coal still available in the ground. Access would continue to be from deep and shallow mines and from opencast workings. Despite the wartime reports of Dott and his colleagues, more proven mining engineers were appointed to re-examine the records. Many of these were available from existing mines or from older sources held by the Institute of Mining in different parts of the United Kingdom and collections housed by the BGS. Deep mining was still very active in western and central Fife at that time, but there was an awareness that only a short way to the east lay the very productive coalfields of the East Neuk. Accordingly, S.B. Forbes, a Chartered Civil Engineer and Mining Engineer, A.M.I.C.E., A.H-W.C.,

A.M.I.M.E., A.M.I.M.M. and Certified Colliery Manager, was appointed to rework any information still available on the coals of the area and to assess the potential for reviving the industry locally within the context of introducing opencast working. In addition to suggesting the sites of estimated residual coal reserves and calculating the tonnages still in place, his thorough review incorporates a huge amount of detail on the former industry. It lists boreholes records and includes plans of many workings. A copy of the report dated 1955 remains in the Library of the BGS in Edinburgh, and copies of a major summary map of part of the district collated by Forbes are held by various residents of Largoward today. Several extracts from this map are included in the report by Terris (2010) and others will be included later in this text. Since 1955 the BGS has returned, drilled exploratory boreholes, published maps, bulletins and a memoir of their findings, confirming the essentially shallow distribution of the coal sequences in eastern Fife, where few of the workable seams reach as deep as 400m below the present land surface. The sequel is a summary derived from these reports, memoirs and academic papers.

The Rires–Balcarres Coalfield

Before attempting to introduce the main coalfield areas, it is useful to examine the historic coal seams worked in the south and southwest of the area. About 1.7km northwest of Colinsburgh a mass of agglomerates from the Largo Law complex forms the high ground of Flagstaff Hill (N56.2319, W2.8851). This ridge extends eastwards past Rires (N56.2308, W2.8693), leading into a 600m-long lowland trough to the south of Sprattyhall and west of the Balcarres Craigs. Between the areas of high ground, the long-since worked out Rires coalfield extended northwards below the grounds of both Sprattyhall and Balcarres. The axis of the shallow synclinal basin in the Limestone Coal Formation is inclined towards the north-north-east, giving a semi-circular outcrop pattern in the south. The succession contained three upper coal seams of varied quality and thickness above the better quality 1.7m-thick *Main* or *Marl Coal*, which included a thin central marl horizon. As Dron (1902) pointed out, the actual position of these coals within the stratigraphic succession was not then known. No limestones were known locally and as access to the coals was no longer available, neither he nor the officers of the BGS could confidently identify the coals. However, the documented history of mining the deposits is of interest.

Although Dron (1902) believed that the coals had been worked out before Landale's account of 1834, the field had evidently not been abandoned. An attempt was made to reactivate it through a newspaper advertisement indicating that the flat-lying main seam at a depth of 42m was in good condition. In 1835 Mr Westwater reported to Mr and Mrs Robert Bayne Dalgleish that Mr Keddie had been recently working an engine pit on the upper coal (the *Marl*

Seam) by room and stoop methods, creating rooms up to 7.5m across, much larger than the agreed 4.3m. He also claimed that Keddie had removed only the large areas of coal which could be easily won by colliers, leaving the more difficult areas for any future incoming tenant, and averred that the underlying Marl Coal in the east had also been extracted in an irregular manner. In subsequent arbitration, Landale reported that his inspection to the base of the workings revealed much waste material within the workings, but none of the rooms seen exceeded 3.7m in width, and they were separated by similarly sized pillars. At the base of the pit the rooms were no more than 2m wide. The oversman indicated that there was one room in the Marl Seam that was 5–6m across. Landale expressed no concern about the stability of the ground above the workings and also recommended that any future mining should use longwall methods of extraction, as they would yield up to 30% more coal for sale. At that time Mr William Westlake had expressed a desire to take on the working of this coal, but his initial proposal to use a hand-operated windlass to drain the workings was rejected.

In a later report of March 1841, Landale advised Dalgleish that Webster had put down a 44m-deep pit and fitted it with a small engine to lift the waters from the Main Coal wastes to the day level. A patch of very good quality coal had been located, but with a 0.2m-thick central band of Marl within it. As the workings extended to the rise (up dip) the Marl thickened to more than 1m towards a bounding fault. The costs of propping the roof and redding the very weak Marl layer were seen to be excessive and, as the areas surrounding the last workings had all been completed, closure was recommended. In a final attempt to keep the coals in production, Keddie explored the grounds of Castle Park, west of Sprattyhall, by means of trial pits, as shown by Michael Martin *et al.* (2007, p.58), but these encountered several dolerite dykes and wastes from earlier workings in all directions. Landale summed up the prospects for the site: 'upon the whole a more unpromising place for mining operations can scarcely be conceived'.

In 1842 it was reported to the Parliamentary Commission on the Employment of Children in Mines that no women or girls worked below ground at Rires colliery at that time but several were still employed on the surface. A fatal accident occurred when one of the women fell down the mine shaft. At that time the now familiar Health and Safety regulations were not in force, and the shaft top was unprotected.

Two kilometres east of Sprattyhall, the Lower Limestone Formation is down-faulted to the west by the Balmakin Fault, bringing northeast-trending outcrops of the St Monance Brecciated Limestone, the *Largoward Splint Coal,* three lesser coals and the Mid Kinniny Limestone to the surface. This succession dips to the northwest beneath the almost horizontal intrusive olivine dolerite sill, which forms the 4km-long eastward extension of the Kilbrackmont Craig–Baldutho intrusion, now displayed in former quarries at (N56.2471, W2.8536) and (N56.2441, W2.7924) respectively.

The intrusion sits beneath the uppermost coal of the Rires basin but above the three lower coals, which dip beneath the sill at about 11° from the horizontal. The younger of these was reputed to have been worked for 'a considerable distance' below the sill at Baldutho Craigs, and there were local claims that a pit had been put down through the sill to work briefly on the Largoward Splint Coal. Landale (1835) recorded that many old disused coal workings lined the foot of the Kilbrackmont Craigs between Baldutho and Colinsburgh.

There is no record of alteration to the coals or limestone where the outcrops disappear beneath the sill to the east of Belliston. However, Landale (1835) indicated that about 200m from Balmakin a 55m thick, east–west-trending dolerite dyke lining a vertical fault greatly impacted the rocks on either side of it. The coals were largely destroyed and the crystalline structure of the limestone was heavily altered, rendering it unfit for use; 'it will not make lime when burned'. The degree of thermal change observed here probably indicates that the hot magmas may have moved along this fracture for a considerable time, allowing baking of the marginal bedrocks to occur. The heating was seen equally on both sides of the dyke, indicating that the crack was approximately vertical when intrusion occurred. It was not necessarily the only pathway along which the intruding magmas travelled.

On either side of the dyke the normally dipping rock structures were also disrupted by many small faults. Beside a similar basalt-filled intrusion near Gibliston House, the coals had been altered and the intervening shales had been 'converted into porcelain jasper with blue lines passing through it'. The remainder of the Lower Limestone Formation outcrop continues north-eastwards to terminate against the Lathones Fault about 2km north of Arncroach. The substantial downthrow to the west displaces the outcrop of the boundary between the Pathhead Formation and the Lower Limestone Formation some 4km to the northwest.

Between the Radernie and Lathones Faults

The base of the Lower Limestone Formation marked by the St Monance Brecciated Limestone outcrops at several locations within the eastern block, principally in the Brewsterwells–Radernie area on either side of the A915 road from St Andrews to Largo. Geikie (1902) described this coalfield as 'a perfectly distinct little coalfield ... strata are disposed in the form of a trough which narrows southward ... and rapidly opens out towards the north'. This, the Radernie syncline, extends beneath the B840 road to Crail from Higham Toll crossroad, and its northward plunging axis meets the Lathones Fault about 250m southeast of the Lathones Inn. The coal-bearing succession, which lies between the St Monance Brecciated Limestone and the Charlestown Main Limestone, is best known from the Higham Borehole (N56.2742, W2.8591). A modified version of the succession quoted by Maxton (1848) is given in Table 6 below.

Table 6: The succession and thicknesses of horizons in the Lower Limestone Formation at Higham, after Maxton (1848) and the BGS Higham Borehole.

Name of Coal/Limestone	Thickness	Quality	Strata Depth
Charlestown Main Limestone	1.5m		Various: 10m
Marl Coal	1.1m	Good	" 2.7m
Main Coal	0.9m	Good soft	" 11m
Duffie (or Duffo) Coal	0.9m	Useless	" 14.6m
St Monance Little Limestone	0.8m		" 0.5m
Brassie Coal	0.5m	Not good	" 11m
St Monance Brecciated Limestone	3.1m		

The outcrops of the St Monance Brecciated Limestone on the two limbs of the Radernie syncline are some 1.5km apart where they meet the bounding Radernie Fault line in the northeast. The outcrop of the limestone of the western fold turns sharply back towards the southwest as it crosses the crest of the adjacent Loanhead anticline, whose core is now occupied by a teschenitic dolerite sill. The westward-dipping limestone is inclined at a shallow angle that allows the *Radernie Duffie* and *Main Coals* to outcrop again at Nether Radernie (N56.2850, W2.8768) where they may have been exploited in the past, but no direct records are known.

Neither the St Monance Little nor the Charlestown Main Limestones are known in this part of the fault block, but outcrops of both the *Largoward Black* and *Largoward Splint Coals* from higher in the succession have been mapped by the BGS, one either side of a second dolerite intrusion at Greigston House. The Largoward Splint was exploited from a pair of pits, the remains of whose shaft tops can be detected in the fields at (N56.2783, W2.8840) on the B941 road 1.2km north of Peat Inn. It is believed that they were worked between 1842 and 1870 respectively by Williamson & Co and Thomas Lumsden, but no records of the workings are known. The northward-dipping Largoward Splint reappears to the north of the Baldinnie–Callange dolerite sill where it dips beneath the Mid-Kinniny Limestone and the *Marl Coal,* which mark the base of the Limestone Coal Formation at Coaltown of Callange.

The thicknesses in the succession of rocks of the coalfield in the Radernie syncline are the greatest recorded within the lower part of the Lower Limestone

Formation in the East Neuk. This is particularly well shown by Forsyth and Chisholm (1968, 1977) who plotted the south–north variations in the thickness of the rocks accumulated during the two complete cycles of deposition between the St Monance Brecciated Limestone and the Charlestown Main Limestone. If each marker horizon was deposited simultaneously across the East Neuk, it is readily seen that sedimentation rates between the periods of limestone deposition differed from site to site; see Table 7 below. It is not clear whether the variation was the result of locally increased sediment input by nearby rivers or whether the central part of the basin underwent enhanced subsidence.

Table 7: The thickness of sediments deposited between pairs of limestones on a north–south section across the East Neuk.

[South]		St Monans Coast	Higham	Winthank	Drumcarro	[North]
	Charlestown Main Lst.					
		15m	30m	14m	14m	
	St Monance Little Lst.					
		18m	26m	24m	18m	
	St Monance Brecciated Limestone					

The poor quality of the *Duffie (Duffo) Coal* at Higham may be linked to the local environmental conditions during deposition, which also led to the contemporaneous development of blackband limestones between Denhead, Winthank and Cassindonald (2.5–4km to the north). The lower *Brassie Coal,* 0.9m in thickness, contained a thin band of shaly sandstone. It was worked mainly in the early nineteenth century, but not in all of the pits. The overlying Duffie Coal was of similar thickness and it also contained a 0.2–0.4m stone band and the quality of the coal was not good. By contrast the 1m thick *Main Coal* was of good quality and sought after for multiple purposes, and the uppermost of the four coals was the 1.3m-thick *Marl Coal,* 9–11m below the Charlestown Main Limestone. It was a cherry coal very popular for domestic use. The upper two coals each rest upon thick seatearths. Each of the worked seams was broken through, and often burned by thin northwest-trending dolerite dykes. The history of working the coals in the Radernie–Higham region was probably quite long, as many of the younger workings encountered old wastes at depth.

Reference to the coloured Google Earth satellite photographs of the area around the Higham crossroad (Figure 8.4) reveals that, in the grounds of Brewsterwells Farm in the field to the southeast of the road junction, the crop marks show many adjacent black circles, often with paler centres. The red

Massey-Ferguson tractor pulling a harrow in the same field gives an excellent scale suggesting that the circles, no more than 4m in diameter, mark the tops of long-abandoned shafts of bell pits. The circles are aligned in a north-north-easterly direction across the field where the BGS 1:50,000 scale map shows the coal outcrop patterns following a similar direction across the field to the east of the lines of the suggested pit tops. The use of bell pits for coal extraction was an early form of mining characteristic of the sixteenth and seventeenth centuries. Unfortunately, no direct records of the early operations in the Brewsterwells–Radernie area have been located.

The introductory report on the Radernie coalfield by Maxton (1848) provided a guide that remains valid today. However, Maxton had much less information on which to base his report than has since been found by borehole and shaft explorations. His final comments were distinctly unenthusiastic: 'It appears highly probable that no extended, continuous or valuable coalfield will ever be found here ... there seems little prospect of the coal of Radernie being a source of revenue ...' However, Geddes (1854) was more optimistic, having established that, significantly, at the base of the *Main Coal* there were Black Band Ironstone balls with a high metal content. He suggested extracting this material, having it tested and possibly taken over by Messrs Bell, operators of the ironstone workings at nearby Winthank for part refining before transport to blast furnaces near Newcastle, with the agreement of the owner, Mrs Bethune.

The Radernie Mine, opened in the early 1870s by Thomas Lumsden (who later developed the nearby Lathones Mines last owned and operated by Montgomery), was the last and most recently working mine in the East Neuk. According to Terris (2010) the mines accessed by a north-eastward drift directly from the surface and inclined at 1 in 14 (about 4°), extended for 475m at their furthest point from the surface. The first and lowermost seam encountered in the drift was the *Brassie Coal*, 10m above which was the *Duffie Coal*, followed for 80m downward and worked by room and stoop methods. The *Main Coal* and *Marl Coals* above the Duffie were worked by longwall methods. The Main Coal thinned from 1m to 0.45m down the dip. Finally, the *Lower Marl Coal* consisted of two 0.5m-thick coal seams separated by a 0.5m-thick stone layer. The working faces of both seams were around 180m long (Terris, op cit.), and these were subdivided into 12.5m sections for each collier. The faces were undercut using an Anderson-Boyes 14-inch coal cutter and after blasting the coals were loaded on a chain face conveyer, whence to hutches to be hauled up the slope to the surface.

Copies of part of a collection of maps from the Scottish Mineral Valuation Office mining plans, showing the evolving patterns of specific coal seam extractions at Radernie, are held in the library of the Institute of Materials, Minerals and Mining at Neville Hall in Newcastle-upon-Tyne. These extracts from Fifeshire Sheets 14, 20 and 21 provide details of workings in the Duffie, Main and Marl Coals (Fig. 8.5) at Radernie. These show that the main drift mine followed the direction of the synclinal axis. The lowermost worked seam, the Duffie Coal, fell from 4651ft. (1418m) at the mine entrance to 4428ft (1350m) above the Forth Coalfields datum some 250m from the entrance. The restricted extractions from the Duffie Coal contrast with those from both of the overlying seams. The longwall workings were often terminated along west-northwest-trending lines, thought to have been defined by dyke intrusions at most localities.

Although the final workings closed in 1946, the closing down sale was not held until 1948, full details of which are in the National Archives of Scotland, Edinburgh. Terris indicated that, in all, the four Radernie coals were extracted from a total area of 6.5 hectares. The roof was supported by Scots Fir props coupled with cross-member timbers 'set like goal posts'. The workings were wet with acidic waters that attacked any metals. It is said that the miners' boots fell apart after no more than two weeks if not washed nightly.

The Lathones mines, situated near the Lathones Inn to the south of the Crail road, are known to have operated successfully during the late 1800s, continuing producing coals until 1920. The extracted coals were distributed locally and also taken by lorries to Largoward Railway sidings for transport elsewhere.

Between the Lathones and Cassingray Faults

The line of the Lathones Fault passes through South Callange, Crossgates, Drumhead, Loanhead, Lathones Inn grounds, Cassingray, North Baldutho and Gillingshill, south of Kellie Law. The south-westward downthrow on this fault is estimated by the BGS to be 91m at Lawhead increasing to 274m at Lathones. Similarly, on the Cadger's Bridge Fault the south-westerly downthrow increases towards the south-east. The area between the Radernie Fault and the Cadger's Bridge Fault is underlain by a succession of rocks leading from the Pathhead Formation in the south-east through the entire Lower Limestone Formation. As in the previous section, the rocks are folded into shallow anticlines and synclines, which have been intruded by several dolerite intrusions during and after faulting. There is a small cluster of volcanic necks around the Bungs of Cassingray.

In the south-east the sedimentary rocks of the Baldutho syncline are overlain by the Kilbrackmont–Baldutho olivine-dolerite sill. Covering the basal St Monance Brecciated Limestone, formerly worked in a series of quarries at Baldutho, the early part of the succession is thinner than at Pittenweem, with the lower coal seams up to and including the either very thin or totally absent *Largoward Black Coal*. Even the *Largoward Splint* seam was recorded as being only 0.46–0.61m in thickness in the mine at neighbouring Belliston. The overlying Mid-Kinniny Limestone outcrops lie parallel to and approximately 100m from the steep southern margin of the sill below which the limestone dips towards the northwest.

The area between East Cassingray and North Baldutho lies southeast of the axis of the Baldutho syncline and is partly obscured by the dolerite sill. The rocks of the nearly symmetrical syncline, whose limbs dip at 9–11°, were extensively explored with shallow boreholes in the early 1900s. Several of the boreholes reached the valuable *Largoward Splint Coal* but deeper penetration towards the less popular *Largoward Black Coal* normally terminated in further thick dolerite intrusions. On the southeast limb of the Cassingray anticline the deeper Cassingray No. 1 Borehole (N56.2588, W2.8490) between East Cassingray and South Cassingray logged the Mid-Kinniny Limestone at depths 21.9m, dolerite to 75m, the Largoward Splint Coal at 76m, the Largoward Black Coal at 99.1m, and the Charlestown Main Limestone at 171.5m. This suggests that both coals and the Charlestown Main Limestone are likely to be present at greater depths beneath the southeastern limb of the Baldutho syncline.

The adjacent Cassingray syncline is an open structure whose northeast to southwest-trending axis plunges at a low angle towards the southwest and can be traced from north of the Lathones Fault to beyond the Cordies Mealling Fault into Lathallan Den. To the south in Balcarres Ward, the fold axis is displaced 400m to the northwest along the Cassingray Fault but has not been displaced further north by the Cadger's Bridge Fault, which terminates beside the Largoward School on the A915 road. South of the Lathones Fault, the oldest coal recorded at the surface on the anticlinal axis is the Largoward Splint Coal, exposed to the east of Bungs of Cassingray. The engine pit beside Bungs penetrated 27.7m to the 1.2m *Largoward Thick Coal*, below which at 53.3m lay the Largoward Black Coal (Telfer 1839). The latter is the oldest seam that cuts across the fold axis 240m to the south of the Cassingray Fault.

In both the Baldutho syncline in the east and the Lathones–North Bowhead syncline to the west, the oldest unit seen, the Largoward Splint Coal, is overlain by the Mid-Kinniny Limestone and the *Marl Coal* (marking the base of the Limestone Coal Formation), and above it the overlying Largoward Thick Coal is encountered. To the east the two synclines become hidden beneath more dolerite sills. To the west and north of the Cadger's Bridge Fault the western limb of the syncline rises towards the overlying Lathones dolerite sheet. However, south of the Largoward Fault a similar rising limb is curtailed and replaced by a further small anticline in the centre of the village. This anticline is terminated along the Largoward Fault, to the southwest of which the Carboniferous rocks exposed at the surface are largely restricted to the gently dipping Limestone Coal Formation and substantial dolerite sills between Gilston House and South Bowhill. The outcrop patterns show both the basal Marl Coal and the younger Largoward Thick Coal to be arranged in open folds, the axis of the easternmost apparently in continuation of the Cassingray anticline discussed above, but now south of the Largoward Fault.

To the northwest of the Gilston block, between the Branxton and Cordies Mealling Faults, a 5km-long and 2km-wide belt of rocks of the Lower Limestone Formation extends from Annfield (in the west) to Falfield, north of Backmuir of New Gilston. North of the Cassingray Fault line the southeastward-dipping succession is weakly folded with dips of up to 15°. South of the fault there is a northward-plunging shallow syncline at Falfield. To the west the Largoward Black Coal, the Largoward Splint Coal and the Mid-Kinniny Limestone outcrops shown on the geological map indicate that the rocks have turned to dip southeastwards. Beyond the Teasses Fault two small structural basins are present on Norrie's Law and to the south of Woodside. At the centre of each of these basins the Limestone Coal Formation is recognized with the *Marl Coal* or the *Appleton Ell Coal* marking its base. The western limit of the block is marked by the Windygates–Hall Teasses quartz-dolerite intrusion, which is terminated in the north by the Cassingray Fault. A second quartz-dolerite sill traces a sinusoidal path southwest of Backmuir of New Gilston. Both of these sills cross the Teasses Fault unbroken, but further north the quartz-dolerite intrusions pre-date the late stage of faulting. In the northwest, between the Cassingray and Lathones Faults, the rocks generally dip eastwards at low angles with no more than minor faulting. There is a continuous sequence from the upper part of the Pathhead Formation through the entire Lower Limestone Formation and into the lower part of the Limestone Coal Formation in the Bandirran–Gathercauld–South Callange area. To the north of Craighall Den the beds dip towards the north and steepen as they approach and are terminated by the major Ceres Fault.

Throughout the general Largoward district there are perhaps only four coal seams of real economic importance, and these are readily identified despite being folded, faulted and intruded by several phases of igneous activity. The coals are normally inclined at low angles and tend to maintain their compositional characteristics throughout the area. The seams most commonly worked were the *Teasses Main*, the *Largoward Black, Largoward Splint* and the *Marl Coal*. In the east the *Radernie Coals,* which were locally significant, were profitably worked for many years.

Collieries of the Largoward district

It would be an overwhelming task to address the characteristics of every pit or shaft in this district, but within this section an attempt will be made to examine the more substantial workings in each of the principal developments. The main features of the southwestern workings in the Rires and Balcarres areas and also those at Radernie have already been discussed in outline, and now attention will be given to the ground to the north of the Branxton Fault and dominated by deposits of the Limestone Coal Group and the two massive intrusions of olivine-dolerite and quartz-dolerite. This is the area north of Lathallan Den including Cordies Mealling, Backfield of Gilston and the village of Largoward to the south-west of the Cassingray Fault.

The coals were being worked in this area during the seventeenth century and early part of the eighteenth century, as the map of Mitchell (1778) indicates not only the active workings of the day but also plots the positions of many earlier sinks, pits and mines on what became known as the Long Mine, (Fig. 8.2). A total of 44 such features were positioned along this structure, a 5km-long day level, which drained the mine waters southward into Lathallan Den (N56.2382, W2.8498) from Lathones in the north through the grounds of Cordies Mealling, Largoward, Cassingray, Louise Hall and the Lathallan estate. The term 'Largoward' was noted by Maxton (1848) as being rather confusing. He indicated that many of the properties were directly adjacent, with the result that it is common to find that one property appears to be blessed with optional and interchangeable names, and the term 'Largoward' tends to be used in a generic sense in many of the reports examined. In relation to the Long Mine day level, the depths quoted from the Mitchell map by Maxton (1848) were said to have been up to 37m below the surface, a figure disputed by Forbes (1955). No actual depths were indicated on the copy of the Mitchell map seen by the present writer. More than 80 further sinks or pits of unknown depth were plotted on the Mitchell map, usually in clusters along the outcrops of individually identified coal seams. The coals sequentially listed by Maxton (1848), some now bearing the name 'Largoward' to aid clarification, are given in Table 8 below. His term 'cubical' refers to cherry coal.

Such a comprehensive series of coals, albeit in small workings, would have taken some years to develop, and obtaining agreements to link sites to form the Long Mine would have required a great deal of collaboration between adjacent landowners. There is a note added to the version of the map seen by Forbes to indicate that two pits by the boundary of Cordies Bank and Falfield had been put down 30 years before, but unfortunately it does not indicate the date that it was 'before'. Perhaps it dates from 1748, but the comment is in a different ink and appears to be a late addition to the map.

Table 8: The Coal sequence suggested by Maxton (1848).

	Coals and Other Strata	Coal Thickness	Strata Thickness	Total Thickness
1	Largoward Parrot Coal	0.6m		0.6m
	Various strata		12.8m	13.4m
2	Largoward Thick Coal	4.0m		17.4m
	Various strata		32.9m	50.3m
3	Marl Coal: Cubical: 0.30m Splint: 0.41m Cubical: 0.71m	1.42m		51.72m
	Various strata		27.4m	79.12m
4	Whitemyre Coal	0.91m		80.03m
	Various strata		58.52m	138.55m
5	Largoward Splint Coal	1.22m		139.77m
	Various strata		20.12m	159.89m
6	Largoward Black Coal	0.91m		160.8m
	Various strata		7.32m	168.52m
7	Lower or Pelcam (Pilkem) Coal	0.6m		169.12m
	Totals	**9.6m**	**159.52m**	**169.12m**

Presumably the early workings were at least partly along the outcropping seams located and proved by the ground breakers. There would have been inevitable conflicts of interest some years earlier as the mineral rights were held by the estate lands. Any extraction of coals to the day level produced by workings from outcrop level or by room and stoop operations at shallow depth could have led to disruption of the land surface being tilled, as already referred to in the disputes between members of the Anstruther family in the area of the Pittenweem coalfield. Much of the later extraction was from the many small bell pits scattered through the area.

A second drainage system was created about 0.5km east of the Largoward crossroads when a wind pump was introduced some 135m to the south of Blakeyhill farm and 335m northwest of the axis of the Cassingray Anticline, (N56.2598, W2.8522). This enabled the working of the coals on the northern part of the western limb of the Cassingray Anticline. Two later steam-powered pumps gave increased access to the northern coals and to some of the coals to the west.

The earliest report seen on the coalfield activities is that of Telfer (1839) who referred to the ongoing workings on the Largoward Splint Coal at Farm, to operations in mines further south, and to both the Largoward Splint and Black Coals as having been abandoned many years previously.

Forbes (1955) summarized many records from the Largoward coalfield, which will guide much of the following discussion. The Largobeath–Cassingray-Balcarres workings spread across a broad area to the south of the Lathones Fault extending across the probable eastern end of the short Largoward Fault to reach the Cordies–Mealling Fault line south of the former railway. Initially attention will be given to the area southeast of the axis of the Cassingray Anticline from a north–south line from Symington's to Ure's Pits and 150m south of the former railway line (Fig. 11.1). At this point it is necessary to stress that, as Michael Martin and Webmasters noted in 2014, over at least two centuries the names assigned to the workings of the Largobeath and Cassingray collieries have gradually become interchangeable, with pits of Largobeath located on the lands of Cassingray and vice versa. The resulting confusion is not readily resolved.

The land surface is inclined southwards, falling from 159.6m in the northwest to 104.2m in the southeast before rising again beside the volcanic neck of Gunner's Law. The Forbes compilation map shows interpreted contours for surface of the Largoward Splint Coal constructed from known depths in the coal workings and from the many boreholes. It indicates that the Largoward Splint Coal dips southwards from 152m above Ordnance Datum (OD) at its outcrop (close to West Cassingray Farm) to reach the level of 0m OD in the southern Ure's Pit, giving a south-southeastward dip of between 1 in 4 and 1 in 5 for the coals in this part of the structure (a figure confirmed by a recorded dip between the Thomson's and Webster's Engine Pits). The reconstruction suggests that to the northeast of a line through the Symington's and Ure's Pits a small syncline, no more than 150m broad, is present beside the Cassingray Fault. Within the area around the base of Webster's Engine Pit, detail from the coal workings reveals five small faults trending north-northwest to south-southeast, each showing a small (3m to 9m) eastward downthrow. To the east of Symington's Nos. 1 and 2 Pits a further four similar faults are seen with eastward downthrows of 2m to 4m. This latter group of fractures is arranged nearly parallel to the adjacent Cassingray Fault, whereas those west of the pits diverge from the main fault at angles increasing westward from 25° to 40°.

What of the coals in these workings? The succession, estimated from the First and Second Statistical Accounts of Carnbee Parish (1793 and 1844) or determined from borehole and pit shaft data, is given in Table 9 below.

The Largobeath and Cassingray Colliery workings presented a relatively uncomplicated area for exploitation. With the axis of the Cassingray Anticline plunging southwestward at a steeper slope than that of the land surface, the oldest part of the succession is to be found in the northeast immediately adjacent to the Cassingray Fault, and the youngest in the southwest. Only three of the coals

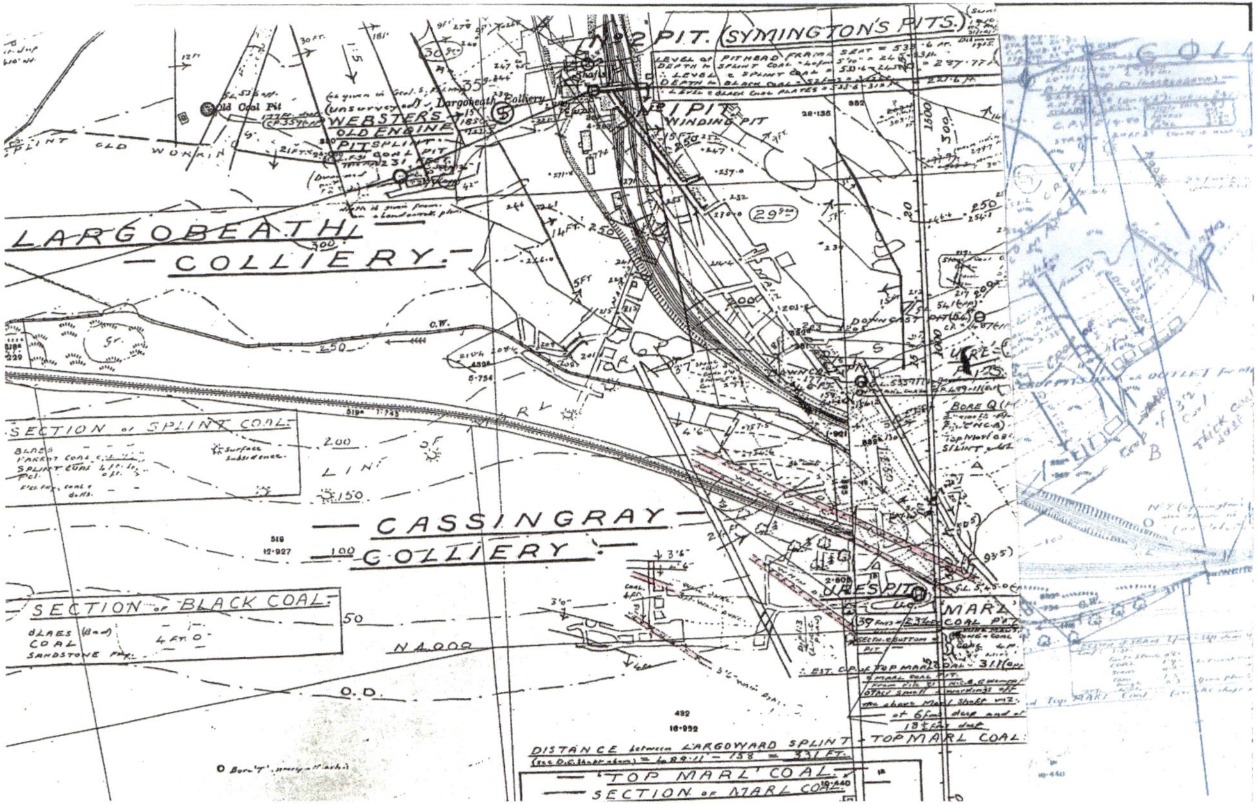

Figure 11.1 Part of the Forbes (1955) collation map of the geology and coal workings south-east of the Cassingray Anticline axis.

Table 9: The coal workings of the Cassingray anticlinal axis, including the Ure and Symington Pits and the Central Fife Railway.

1	Top Marl Coal	1.4m	
	Strata	16.0m	
2	Splint Coal	0.4m	Marginally thin for working
	Strata	69.2m	
3	Largoward Splint Coal	1.5m	Best quality coals Parrot (0.19m) Splint (1.32m)
	Strata	20.2m	
4	Largoward Black Coal	1.2m	Adequate (but Largoward Splint preferred)
	Strata	1.0m	
5	Scarrat Loft Coal	0.6m	Not greatly worked
	Strata	8.0m	
6	Thin Splint Coal	0.5m	Not greatly worked

Table 10: Sections of the three main coals at Cassingray as recorded by Forbes (1955):

Marl Coal

Roof	Blaes (weak shales)	
	Coal	0.1m
	Fireclay	0.38m
	Hard Rib	0.13m
	Fireclay	0.86m
	Coal and Stone	0.05m
	Coal	1.2m
Pavement	Sandstone	

Largoward Splint Coal

Roof	Blaes (weak shales)	
	Parrot Coal	0.2m
	Splint Coal	1.3m
	Fireclay	0.01m
Pavement	Sandy Fireclay, Coal and nodules	

Largoward Black Coal

Roof	Blaes (weak shales0	
	Coal	1.2m
Pavement	Muddy Sandstone	

had long histories of exploitation (Table 10 above). The lowest and geologically earliest of the three was the *Black Coal,* which extends below the entire area. The highest was the *Marl Coal,* which first appears at outcrop about 60m north of the late nineteenth century downcast Ure's Pits and extending for at least 150m both west and east of that locality. From here the excavation shaft could have been pushed deeper to encounter the remaining coal succession below Ure's Marl Pit, but there is no evidence that this took place, as those coals (*Scarrat* and *Thin Splint*) were not normally of sufficient value to be worked.

The principal dooks on the *Largoward Black Coal* were taken down the slope of the dip towards the southeast for about 120m before the economic interest led to preferential attention being given to extraction of the overlying and more valuable *Largoward Splint Coal*. The uppermost working shaft was that of Thomson's Engine and Pumping Pit, situated about 110m south of West Cassingray Farm. The land surface elevation at the top of the shaft is recorded as having been at 152m above OD; the Largoward Splint Coal was met 29.2m down the shaft and the Largoward Black Coal at the depth of 51.2m. Only 0.99m to 1.07m in thickness, the Largoward Splint Coal was highly sought after for household use. The same coal was located at a depth of 67m in Webster's Pit, where it too commanded a good price. The *Marl Coal* is believed to occur some 91m below the Splint but no attempt was made to reach it from this location. There was a strong belief that the small-scale faulting was increasing in frequency westwards towards the crest of the Cassingray Anticline.

In 1839 Telfer observed that at Bungs of Cassingray, 150m north of the Cassingray Fault in a position geologically similar to that at Cassingray, the steam-powered 20HP engine at the pit (which had been worked formerly as a horse-powered gin and barrel operation) had become deeply rusted and had fallen into a state of disrepair. Gemmell (1905) attributed this to the presence and decomposition of iron pyrites in the roof of the seam. Pyrite fragments left in the wastes were also thought to have induced fires through spontaneous combustion, and in order to suppress the potential dangers, the dross was removed from the workings and taken to the surface. The pyrite was also blamed for the introduction of acids into the mine waters, leading to destruction of the iron pipes and pumps, as reported earlier at Radernie.

The greatest surviving detail of the workings in the Cassingray–Largobeath mines provided by Forbes using data from Landale, Frew & Gemmell (L., F. & G., 1903) refers to activity in the southeast in an area below the former railway line. The two shafts of the Symington's Pits lie in the north and two Ure's Pits, including the Marl Pit, are to the south. The shaft of the latter passed through two relatively poor coals at 11m (1.07m thick *Largoward Black Coal*) and at 34m (the 0.61m *Largoward Thick Coal*) before reaching the higher quality 1.47m thick *Marl Coal* at a depth of 33.8m. Levels 24m in length were driven in the direction of strike from the shaft in each of these horizons, but coals from only the upper seam were used to power the engine as the shaft cutting continued. The Marl Coal, 1.4m thick at the base of the shaft, consisted entirely of good, rough steam coal with the exception of a thin stone rib that thickened westwards to 0.15m. The seam had a hard sandstone floor but the roof was mainly of fireclay, requiring much shoring by timber and steel girders. By 1903 it had been proved to extend 365m in an east–west direction and 137m northwards before it encountered the former workings of the Ure's northern pit by the foot of the 14m deep ventilation shaft.

The Main Dook, the principal working roadway on the Marl Coal, followed a path bearing 143° between the Symington No. 2 Pit and Ure's Downcast Pit, falling from 300m OD to 156m OD in a distance of 295m. A second similar dook 5m higher and leading directly to the Downcast Pit followed a route 23m to the west and was linked to the Main Dook by at least eight cross-cut routes. The collation map by Forbes (1955) shows side galleries both east and west of the two major dooks leading to substantial areas at least 150m across, which provided structural details compatible with the total extraction normally achieved with longwall operations. The western extensions include Nairn's Dook, a route leading north-northeast to south-southwest from the western main Dook, and this not only led to abandoned longwall workings but also provided an extension leading to the south of the former railway embankment. Both to north and south of the former railway a series of five whinstone dykes was encountered, each little more than one metre in width, but which had nevertheless baked the coals equally on each side of the igneous rock, indicating that they had been intruded vertically through the overlying dipping rocks. These linear structures trending 120°–130° were noted in the quadrant west to northwest of Ure's Marl Pit. The Forbes map indicates that there was some continued working to the west of that mapped by Landale, Frew and Gemmell (1903), and the meeting with many small faults in the workings confirmed the warnings that there was likely to be an increase in faulting in that direction towards the axis of the Cassingray Anticline. However, there were similarly large numbers of small faults with small downthrows to the east between Webster's Pit and the Symington pits and beyond to the east. The alignment of the fault planes is about 30° northeast from that of the nearby Cassingray Fault.

During the operation of the mines, the shafts and workings were penetrating well below the level of groundwaters in the rocks. The extraction and removal of those waters was vital to permit the winning of the coals. In a relatively simple structural arrangement such as characterizes the eastern limb of the Cassingray Anticline, the waters were removed from the working faces largely by the deployment of substantial pumps.

In his 1905 report, Gemmell indicated that the early drainage waters, pumped up to the Largoward Day Level at 37–46m below surface, were discharged westward into the larger stream issuing from the Largoward Day Level that drained most of the western flank of the anticline. Later, pumped from the much deeper mines of the more developed Cassingray coalfield, the waters were carried along a second day level and discharged into a surface drain leading westwards about 60m north of the railway line. After joining the greater flows of water from the Largoward Day Level, the waters passed beneath the railway embankment in a culvert to flow southwestward to Lathallan Den.

Table 11: Output and disposals of coals from Cassingray and Balcarres between 11/11/1903 and 11/11/1904.

Output			**Disposals**	
Stock at start	252 tons 12cwt		Screened coals	4,418 tons
Coal and Dross Cassingray	11,956 tons 5 cwt		Sales of triping	7,563 tons
Coal and Dross Balcarres	3,624 tons 13 cwt		Sales of Dross	2,462 tons
Excess of disposals	1,802 tons 11cwt		Workers Free Coal	260 tons
			Engine Dross	1,560 tons
			Stock 11/11/1904	709 tons
Total		16,973 tons 16cwt		16,973 tons 16cwt

The final phase of development of the Cassingray mines, then owned by the Earl of Lindsay, continued into the period of the First World War. In 1910–1911 Symington's two shafts were put down to the *Largoward Splint Coal* to the east of Webster's former workings as shown on the map. The winding pit reached the *Largoward Black Coal* at 97m and a cross-cut mine was cut south to the Largoward Splint. The second dook (Nairn's) is shown on Forbes's map. Both seams were extracted until the end of 1913, when the working became unprofitable; for example, the roof of the Largoward Black Coal was weak and required too much timber to support it.

As coal production was about to start in April 1911, the west-going level approached Webster's abandoned workings on the Largoward Splint Coal, which had not been accurately surveyed. (An error of positioning by 42.7m was later determined.) At night the water burst through into the new tunnels and up the shaft. Fortunately, no personnel were present. After a month pumping out the waters, the working of the coals was resumed. The inflowing groundwaters continued, but these were readily removed by an air-driven pump at the foot of the shaft and winning of the coals was restarted. The 1.07m thick Black Coal here was largely of cherry, but had a weak roof requiring much support. The cherry coal here was sold for domestic use under the title of 'jewel'. Dott (1944) quoted the daily output from both seams as 115 tonnes.

The significance of the mines to the local economy must not be overlooked. The Largoward area supported an essentially agricultural community with mining as a secondary employer. In 1844 no fewer than 33 colliers, 3 labourers and 2 engine-men were listed as working in the mines. In 1905 Landale, Frew and Gemmell (L., F. & G.) noted 36 miners, 19 drawers and brushers below ground and 12 more on-cost and engine men on the surface. By contrast in 1913 as many as 101 were employed at the mine, 55 people in the Largoward Splint Coal and 10 in the Largoward Black Coal. There were 28 oncost men below ground and 18 on the surface. During the month of September 1913, in a period of 24 working days 2202 tons 9 cwt of coal were taken from the Largoward Splint Coal and 580 tons 8 cwt from the Largoward Black Coal. L., F. & G listed the total output and sales from the Cassingray mines during the final six months of production until May 1914 (Table 12). However, the operating workings were becoming restricted with half of the working faces in Balcarres ground and the other half in Cassingray. L., F. & G. expressed themselves convinced that 'in ordinary times they can do no good working the Marl seam by itself' but they understood that fresh borings would be made to seek the Largoward Splint Coal 90m below the Marl Coal in both Cassingray and Balcarres. However, the start of the Great War 1914–1918 brought considerable problems for the enterprise, as the coal-fired German Navy was one of the major customers.

Table 12: Production and sales of coals from Cassingray and Balcarres between 11/1913 and 5/1914.

Output		**11/1913–5/1914**	**Sales**	
Splint Coal	22,173 tons 15cwt		Splint	11,730 tons 16cwt
Black Coal	3,775 tons 17cwt		Other	5,335 tons
			Trebles	216 tons 12cwt
			Dross	2,322 tons 18cwt
Total	25,949 tons 12cwt			19,605 tons 6cwt
			Workers Free Coal	552 tons 6cwt
			Engine Fuel Coal	6,343 tons 18cwt
Actual Sales				25,949 tons 12cwt

At this stage it is useful to consider what these figures meant to the enterprise in terms of income being generated. A few of the suggested pricing levels included in agreements between the landowners, the Earl of Lindsay within the lease of the Cassingray property to Mr Robert Ure for the mining company, are listed below. The complete set of agreements extending from Martinmas 1899 also included items such as the rental to be paid, royalty to be raised on the products, and any restitution for damages incurred to the properties, including access routeways to the site for removal of the coals. These are listed in Table 13 below. Many of these figures were notional and normally remained flexible according to the success or otherwise of the mine.

Table 13: Royalties on the Cassingray Lease, 1899.

Largoward Splint	8d
All other common coals	6d per ton
Triping	4d per ton
Dross	2d per ton
Cannel Coal	10% of selling price
Shale	6d per ton
Raw Blackband Limestone	6d per ton
Fireclay	2d per ton
Burnt Lime	5d per ton

Restoration of agricultural land: £40 per annum. Not to fill up any pits without consent.

West of the axis of the Cassingray Anticline but between the Cassingray and Cordies Mealling Faults, the sedimentary rocks and coals dip principally towards the northwest and outcrops ranging from the *Largoward Splint Coal* to the *Largoward Thick Coal* outcrop beside the Cassingray Fault itself. To the southwest, within the faulted unit as marked on the BGS map, is the curving outcrop marking the position of a small syncline whose axis points towards the Inn by the crossroads at Largoward.

Lathallan, Gilston and New Gilston

In an account by John Grieve (dated 4th June 1829) of the Rires coalfield underlying Lathallan Park, it was reported that in *c.*1789 some 34 acres (13.8 hectares) of Rires land had been sold by Mr Bayne to Major Lumsdaine under reservation of the subterranean minerals. These 34 acres (13.8ha) of surface became the East Part of Lathallan Park, below which Lumsdaine is said to have mined the four Lathallan coals that Grieve had been led to believe could be worked under this ground, as previously some had been worked northwards in the Lathallan lands. Landale had been unable to learn the quality or thickness of these coals and recommended that a borehole be put down between Lathallan House and the March Burn before any speculative purchase of machinery be undertaken. The borehole confirmed the presence of viable coals, two seams each 1.0m in thickness, the third 0.8m thick, but the fourth too thin to work. All were of good quality splint and later Dron, thinking that the line of shallow pits along the Long Mine would have been put down to relatively young coals, ventured to suggest that these coals might be equated with the Largoward coals (Dott, 1944). Noting that the beds thought to incorporate the coals dipped regularly at about 1 in 6 towards the east, he suggested the cutting of trial pits east of the borehole site in the belief that some of the younger coals might have extended southwards beneath the overlying dolerite sill recognized at Rires Mill. In his defence, Grieve later (1830) reported that this part of the Lathallan Park was originally referred to as 'Firth Mine', presumably deriving the name from a former enterprise beside the B941 road marking the eastern boundary of the estate. It was rumoured, but not confirmed, that Mr Sandeman, the new owner of the policy, was preparing to work the mine for his personal coal supply once a year and storing the coals in his cellars before covering the mine with branches and turf so that it would not detract from the appearance of his policies.

Coals outcrop in the northern and eastern parts of the estate. The 1.2m thick *Largoward Black Coal* was found in the north at the depth of 165m. A later exploratory hole south of Lathallan House located the 1.2m thick *Largoward Splint*. The peak production from these deep seams reached 200 tons per day. (Landale, 1835). Other thin coals were noted in each of the boreholes, as were thin limestones that Geikie (1902) suggested might be the Index Limestone at the base of the Upper Limestone Formation, but which Forsyth and Chisholm (1977) assigned to the Mid-Kinniny Limestone within the Lower Limestone Formation. The outcrop pattern provided by the BGS 1:10,000 scale map suggests that the beds are rising and falling across the hilly ground towards the Largoward Railway Station between the massive dolerite intrusions of Kilbrackmont and Gilston. The *Marl Coal* marking the base of the Limestone Coal Formation is readily identified in two down-faulted blocks towards the southern end of the Cordies Mealling Fault.

Gilston

The earliest known account of the coals on the Gilston estate (N56.2468, W2.8923) is a report by John Geddes dated 23 April 1859, which presented a very general description of the property apparently produced for the sale of the estate at that time.

The Gilston Estate lay entirely within the coalfield area, but much of it is occupied by intrusive dolerite sills that degrade and destroy many of the coal seams. Landale believed that the early maps by the BGS showed more dolerite than was really present. However, the intrusion about Dunikier (Dunnicher) Law certainly extended south and west into Gilston, destroying the coal all around it along the north of the property and halfway into the grounds around the steading of Berryside. This same sheet of dolerite extended east over most of the three northern fields and the southern March fields of South Falfield. Another ridge of this rock on the south March fields of the same farm underlay the gardens, the Mansion house, the farm of West Lathallan and part of Gilston Mains Farm. The rocks at Balhousie were very much troubled and contorted east of the narrow tongue of land, where it was disrupted by the volcanic centre of Largo Law to the south so that any coal seams in that neighbourhood would prove to be unworkable. Only in the ground not invaded by igneous dykes, sills or volcanic necks could the coals be sought realistically.

According to the old colliers, the two pits marked on the first OS map were sunk about the beginning of the nineteenth century to the upper or *Parrot* seam some 25m deep. The workings were drained by a day level, making it water free. However, little coal was ever taken from these pits, the seam having been burned in place to the extent that it had become 'uninflammable'. The colliers added that David Forgan, when Manager for Durham, put down trial pits on the *Marl Coal* near the north lodge but again found the coal troubled, and no further work was done in it.

The *Largoward Splint Coal* was worked adjoining Gilston, the last fitting reaching 116m in depth, which surprised Forbes (1955) who challenged the identification on the basis that the splint was normally about 70m from the surface and the depth claimed was unexpectedly great. It ended on foul burned coal dipping beneath Gilston at about 1 in 5 before being destroyed as usual by the intrusive rocks. Working in it continued until the coal would not burn in the engine furnaces. The Largoward Splint Coal was next seen to the north of the Cassingray Fault, in a basin form on the Falfield Estate, where the workings also ended on burnt coal. In the 84m deep No. 1 Pit at Falfield, which cost £4000, the coal had been burned and was almost useless. The *Poor Parrot*, the thickest part of the seam, won by day level pits up to 46m deep, had been worked extensively many years earlier (1840–50?) for the Parrot itself, which was sold cheaply as a household coaling those days. Landale believed that it had been worked up to the Cordies Mealling Fault, approximately 280m beyond the Gilston North March. The late Thomas Brown put down a 24m deep pit to the *Largoward Thick Coal* close to Gilston but also found it burned and useless.

To the west, where the land slopes continuously from about 200m at Norrie's Law and Teasses to around 100m beside the Keil Burn, the complete succession of the Lower Limestone Formation from the St Monance Brecciated Limestone to the basal *Marl Coal* of the Limestone Coal Formation is seen in northeast to southwest-trending outcrops. The maps of Michael Martin *et al.* (2007) show the presence of outcrop pits uphill from the Boghall Mine. The section of the coals worked on Baldastard and Teasses Common was as follows:

1. The *Buffo Coal* 0.7m thick at 55m beneath varied strata.
2. The *Splint Coal* (in 2 leaves) 0.7m thick at unknown depth. Worked to depth of 80m.

3. The *Main Coal* (in 2 leaves) 0.77m thick at depth of 4.6m of varied rocks. Worked to depth of 110m.
4. The *Under Teasses Coal* 0.66m thick 4.6m below Main Coal.

According to David Bauldy, the Buffo Coal had a strong roof of blaes and a hard pavement, requiring the holing to be done in the coal itself. An uppermost 0.1m layer of soft Cherry coal lay above a stone rib 0.05m thick, beneath which was a 0.25m thick Splinty coal with a basal layer of Rough Coal. This could be readily mined, and a collier could cut 12 loads, nearly two tons a day from this seam.

The *Splint Coal,* which also had a stone layer within it, was of better quality. The stone layer increased in thickness eastward so that only the upper leaf of the coal reaching 0.5m was normally worked, mainly for lime burning. The last 80m deep pit on this coal was wet but was worked from 1873 until 1892. An area of 30 acres (121ha), faulted both up- and down-dip, was excavated. A trial lease was taken by Mr Martin, the Lime Tenant, but this was not continued.

The best seam was the excellent *Main Coal,* which was widely worked down to depths of 110m. The upper leaf of *Splint Coal* was 0.25–0.38m thick and the lower 0.46–0.50m thick layer was of *Cherry Coal.* Together they provided a good house coal, which burnt to give a brown ash. The two leaves were separated by a 0.40m layer of shales. A day's work for one collier produced about 25cwt of coal. The splint layer is sometimes equated with the Dunfermline Splint of West Fife.

Separated from the Main Coal by 4.6m of black shales, the *Under Coal* was about 0.7m thick, including a thin upper layer of gas-free Parrot Coal. Lacking the quality of the overlying Main Coal, the Under Coal was normally worked by longwall methods from a cross-cut mine from the Main Coal, and was largely used for lime burning.

The Main and Under Coals were worked for many years from a 66m-deep engine pit sunk for Lady Mary Lindsay Crawford 500m north of Bonnyton House. These two coals degenerated eastwards until both were useful only for lime burning, for which they were worked. The next stage of development planned was to have been a second nearby engine pit 150m to the southeast to be worked down dip by longwall to 110m depth, but 20–30m north of the Teasses Fault. Following Her Ladyship's death her son, the Earl of Glasgow, leased it to Mr George Grieve, but heavy watering caused problems for the pumps, after which the coals became burnt, foul and troubled about 160m from the Baldastard boundary. Mining activity continued by working the abandoned lime coal pillars until Graham and Patten were said to have fitted a new working on the *Splint Coal* at Balmain 2km south of Baldastard Mains, in an area south of the east–west trending Branxton Fault that defines the southern boundary of the area under examination.

The extended workings on the *Main Coal* that had been traced to a depth of 100m were abandoned, and shallow workings, mainly shallower than 30m, were developed to extract coals for lime burning using horse gins and windlasses. The working was water-free to a depth of 33m. The day level discharged into the Keil Burn and flowed away through Pitcruvie and into the Forth at Lower Largo. Landale (1890) recorded that when the engine was working on Graham and Patten's pit at Balmain, the waters ceased to flow from the day level at Baldastard, although there was no evidence to show any direct connection between the mine and the day level, which also successfully drained many small workings in the Baldastard Plantation.

Shallow gin and windlass workings drained other pits on the Baldastard part of the Teasses Common lands, exploiting what Bauldy reported as a 0.45–0.50m-thick *Cherry Coal* ribbed with splint some 90m above the *Main Coal.* A further 33m above the Main Coal was a 0.67m-thick rough coal known as '*Blue Eye*', which lay beneath a strong freestone roof. This good-quality coal was usually worked longwall. Landale (1879) recorded that Mr Martin had a contract to raise the coal for 3s 0d per ton on the hill, but at that time the sales price for the coals was higher than charged for similar coals brought by rail from the west of Scotland. The prepared workings were effectively mothballed with several walls ready to be worked when economic conditions improved sufficiently to permit a profit to be made on the coal sales. There is no record of their having been subsequently brought into operation.

The *Splint Coal,* worked by an engine at Bonnyton, extended east towards Gilston but deteriorated in quality and thinned in that direction. Extraction ceased when the seam narrowed to 0.5m.

Mr Cochrane, the borer, took a lease of the coals on Gilston Mains from Mr Baxter. When he opened up the old wastes he entered the coal above the Boghall Ingress (N56.2565, W2.9248) marked on the old OS map. He installed a small engine to work pumps with the intention of emptying the old wastes. However, neither the pump nor the engine was sufficiently powerful. He recruited two partners to form the East Fife Coal Company, and installed a much larger engine and pumps able to lift 400 litres a minute. After lengthy pumping the works were emptied, to discover that the mine had been drained previously by an old level from Baldastard draining the Gilston coals to a depth of 25m. The roof was 0.8m of good sandstone, above which was a 2.1m-thick limestone. The waste was upstanding with rooms 4.6–4.9m wide separated by pillars 1.8–3m square. The coal was 1.14m thick, the lower 0.2m near the pavement was poor and sulphurous, but the overlying *Cherry coal* was of excellent quality and attracted a good price in the market. However, the works encountered a considerable amount of trouble in the form of dolerite intrusions. Despite putting down many boreholes, no space was found large enough for a deep pit to be created, and this led to the cutting of many shallow pits to remove the coals. When recalled to reassess the mineral prospects for the Estate, Landale (1879) lamented that he had been unable to locate anyone who had worked in any of the mines on Baldastard and so he lacked fresh first-hand accounts to include in his report. The works were abandoned by the Gilston Estate.

That was not the end of the coal working in this area, however, as the new owner, the Earl of Elgin and Kincardine, let the lease to Mr Archibald Whittaker for 20 years from 1896. At the outset, a pit 50m south of Appleton, nearly a kilometre east of Baldastard, was working the flat-lying *Ell Coal* 27m below the surface. The 0.53m-thick coal was suitable for household use, but it had a thick, soft blaes roof that was difficult to support in both workings and roadways. Much of the blaes was carried to the surface in the hope of using it in brick-making. The seam rose to the north and thinned to 0.46m, which was too thin to work, and the site was abandoned in July 1898.

A new pit beside the East Fife Central Railway line was put down in 1898 seeking the 0.8m-thick *Largoward Splint* (then referred to as '*Blue Eye*') at 58m and the underlying 0.48m thick *Largoward Black Coal* at 76m, as discovered by borehole investigation. The pit sinking revealed disturbed rocks, heavily faulted, which carried large volumes of water. Installed pumps failed to keep up with the inflows as fine materials from the roof clogged the pump intakes. A second pit driven to 32m encountered very little water, which was easily handled by a moderately sized pump. However, this pump was stopped in May 1900 due to lack of funding. The workings had been equipped with winding and pumping engines, with three boilers and a chimney stack erected near the brickworks, with fully equipped buildings and workshops and a siding leading to the railway. Despite the expenditure of a considerable amount of money on drilling additional exploratory boreholes, the locations and interrelationships between the coal seams nearby and further afield were not clear due to faulting and the burning of the rocks in contact with the igneous intrusions. The lessees were several years behind with paying their annual rent, to which they were contractually bound until at least 1906.

The *Buffo* and *Teasses Main Coal* workings ended in burnt coal, but the disappointed Landale suggested that there might be 'an immense' tract of coal unworked extending north through Woodside and on towards Ceres. Later exploration for the possible coals went unrewarded, as only thin coals were encountered in boreholes. No spectacular reserves of coal were located.

New Gilston

In 1879 Messrs Beveridge and Darling asked Landale to report whether their quarry and its minerals could be turned to better account. Landale visited the quarry with George Strachan, who had worked in the quarry for more than a decade, and went over the ground with him. No new discovery had been made since Landale's previous visit in 1849, which had led to the issuing of a trial lease on the 'rums' layer to a Mr Galt, who made some borings but did not enter into a lease. Extracts from some of the Journal logs concerned are provided below. Beneath the *Blind Coal* at the bottom of the quarry lay shales that might have included some oil-shale, which Galt had not bored. He put down thirteen bores, six of them to the coal and Parrot lying between the outcrops of the igneous sill and the village.

Journal 1: South of the sill at the edge of the first fence east from the quarry road, the 60-70m-deep bore found both the Parrot and the Cherry Coal which was with it, all undisturbed at 24m depth.

Journal 2: About 73m east from the first. The same coal seam in the same state at 20m depth.

Journal 3: Site 50 paces (45m) south of the last bore. Coal at 22m depth and the bore continued 5.5m into the Rums of the quarry and ended in dolerite into which it bored 0.5m.

Journal 4: A little further east and about 180m north from the middle of the village. Between 20 and 22m deep to the same coal, proving the considerable extent of the seam lying regularly but damaged by the intrusion.

Journal 13: East of the mouth of the Day Level where the sandstone cropped out irregularly and dipped moderately towards the quarry. Deepest bore, penetrated 62m in bedded rocks and ended in dolerite into which it drilled 0.3m before stopping. 0.1m of common coal located.

Table 14: Section through the coals of Peattieshill based on borings by Galt.

Coal	The Peattieshill seam	0.98m	0.98m
Strata	Various rock types	73-93m	93.98m
Splint Coal	Good quality coal	0.51m	94.49m
Strata	Various rock types	45m	139.49m
Cherry Coal	Coarse	0.23m	139.72m
Parrot Coal	Coarse	0.2m	139.92m
Parrot Coal	Good	0.33m	140.25m
Parrot Coal	Coarse	0.25m	140.50m
Sklut	Hard micaceous black shale	0.2m	140.70m
Bituminous Shale or Rums	Hard, sandy and inferior	3.0m	143.70m
Main Rums	Good	6.4m	150.1m
Coal	Coarse and partly blind	0.9m	151.0m
Black Bituminous Shale	Said to have been bored through	29.3m	180.3m

During the nineteenth century two locally used terms, 'Rums' and 'Luncarts' were applied to irregular, apparently disturbed and internally structureless, slightly bituminous black shales. Neither of these terms survives today in common usage. Both may have had similar origins, the rums generally being smaller than the luncarts. Geikie (1902, p.177) referred to a 30m layer of Rums being worked in a quarry beside the Clockmadron to New Gilston road (N56.2673, W2.9235) where it was overlain by the 1.2m-thick, long-abandoned and heavily wasted *Peattieshill Coal* seam.

Galt, who had been in search of oil shale or bituminous coal, drilled only direct towards the coarse Parrot Coal that overlies the Rums Quarry. Strachan reported that Galt had tested all the Rums in the quarry for oil and found them to yield only about I gallon (4.5 litres) of oil per ton, with the exception of one yellowish stratum 4 inches (0.1m) thick, which yielded 16 gallons (60 litres) of oil per ton, but of poor quality. Nothing less than 20 gallons (90 litres) per ton and of good quality and low specific gravity oil would have paid. None of these Rums were suitable for making oil.

The origin of the word 'Rums' is not known. According to Barrowman (1886) they consist of apparent upwellings of hardened shale of regular 'semi-circular' shape with diameters of up to 5m and circumferences rarely to 10m. Most were of smaller size, and they lacked formal internal structure, but there were often traces of loam between the shale masses. The coal seam at the bottom of the Rums quarry was blind and next to useless. Most of the strata here and in New Gilston village had been similarly damaged by the underlying and now outcropping igneous rock.

The 5.0m–5.5m-deep Rums quarry had been worked slowly but regularly by Strachan and one helper over a period of many years. The base of the Rums 'lay flattish like the surface of the ground'. The deposit was worked as a low-quality fuel under a verbal agreement for a final rent of £45 per year and a royalty paid at half the sales price. Often £90 worth was sold in a year at a price of 1s 1d per cart load of 13–14cwt, or about 1s 1d per ton. This left the quarryman about 10d per ton for tirring and quarrying the Rums. The annual sales varied between £90 and £130.

Both the *Splint* and *Parrot Coals* appear to have ended at a dolerite dyke. The Splint Coal, worked by an engine at Bonnyton, extended east towards Gilston but deteriorated in quality and thinned in that direction. Extraction ceased when the seam narrowed to 0.5m. To the east of the dyke and the track to Clockmadron, (N56.2673, W2.9235) the seams were foul and badly troubled. Pryde attempted to approach the area from the east by means of a cross-cut mine from his workings at a depth of 22m below the surface, but again encountered bad conditions and recommended that no further money be spent on exploring that area.

Landale concluded that there was very little hope of any common coal being found workable on the Gilston estate, the only chance being of the Parrot Coal at Gilston Mains turning out to be a gas coal of value. Should the nearby Baldastard tenants find it in the future, he suggested that Mr Baxter should make a few more bores and trials and if found, advertise for a tenant. No further progress is known.

A detached luncart of gas coal had been recognized near Gilston Mains with loam above and below it. Without presenting further evidence, Landale claimed that it was 'evidently water-borne' and locally sourced. However, since it had neither any defined dip nor internal stratification within or near it, the claim of generation by water must be challenged today.

According to the Scottish National Dictionary of 1700, luncarts were patches of structureless coals encountered at Baldastard and Gilston. The word was thought to have been used for a large lenticular mass or nodule of one mineral in the layers of another. In many instances what the quarrymen also referred to as 'lunearts' (*sic*) occurred in solid, upward-broadening cone-shaped masses of very large dimensions, unstratified and without cleavage when freshly exposed in any direction. This coal was said to have been overlain by blackband ironstone occasionally rising to 6 inches (0.46m) in thickness in 'lunkers'.

Landale burned samples of the material from the luncarts in various ways and found that they burnt like a gas coal. He concluded tfrom its general properties that it was more like a Parrot Coal than the surrounding shales, and deduced that its site of origin could not be far distant and was probably a displaced slice from the uppermost Parrot Coal in the typical underlying strata. Very similar but smaller structures had been reported at Baldastard, where the workers found it repeatedly in very disturbed condition, sometimes in cartloads lying in irregular shapes and orientations, sometimes horizontal, elsewhere vertical and always broken and occasionally described as having a conical shape broadening upwards. It was always surrounded by loamy earth, none of which was bedded within the strata. Many unsuccessful trials were made to find its source.

Today doubt would be cast on the 'water-borne' assertion, and an alternative suggestion would refer to disruption by either water or gas escape phenomena, or possibly to seismic activity shaking the sediments before they had become consolidated. Today the disturbance is more likely to be attributed to the action of gases having been released from, and rising above, the gas-rich coals as they were heated and partly burnt by intruding igneous rocks below the seams.

It appears that the original oil and bituminous matter had been largely driven off in vapour, which disrupted, impregnated and enriched the overlying shale with bituminous fluid to form the impure fuel for which it was often used locally. The heating had been supplied by the molten basic magmas injected into the layered coal and shale sequences in late Carboniferous times.

For the purposes of this analysis, it will be most useful to continue an examination of the rocks outcropping between the Branxton and Cassingray Faults, to the west of a north–south-trending fault along the line between Woodside to Balhousie and separating the New Gilston area from Baldastard, Bonnyton and Teasses. Although the coals were affected by the igneous intrusions, their destruction was less substantial, and successful collieries were established to exploit the older coals of the Lower Limestone Formation to the west of the fault.

Chapter 12

Coals of the Lower Limestone Formation around Largoward

The best section for the extraction of coal was towards the east end of the property at Largoward, where the well-known seams of coal of the upper part of the Lower Limestone Formation outcrop (Table 15).

Table 15: The coals of the Lower Limestone Formation at Largoward.

1st	The Parrot Coal	0.22m of Black Coal a cross-cut mine from the Main Seam with coal at base	0.76m
	Various strata		9.1m
2nd	The Largoward Thick Coal	Sometimes as thick as	4m
	Various strata		42m
3rd	The Marl Coal		1.02m
	Various strata		12.8m
4th	Thin Splint Coal		0.41m
	Various strata		14.6m
5th	The Whitemyre Coal		0.76m
	Various strata	Containing Two-Foot Limestone	58.5m
6th	The Main Splint Coal	Largoward Splint (Blue Eye Coal)	1.07m
	Various Strata		22.0m
7th	Largoward Black Coal		1.07m
	Various strata		6.4m
8th	The Pilkem Coal		0.46m

The 1st or *Parrot Coal* seam was examined in Largoward by Thomas Brown, where it was 0.76m thick with a layer of 0.2m to 0.23m of Black Ironstone, which sold for 13/- per calcined ton on the hill. The costs of carting the Parrot Coal to the station, and railway dues diminished its value to the producer and it did not sell well, although offered at 2s 3d per ton less than the *Splint*. The *Marl Coal* (3rd) was hard when first exposed but soon crumbled to 'marl' and had to be thickly set with props. A collier could put out only 24cwt each day, which was not economic, and the pit was abandoned in the autumn of 1884 after considerable loss. In previous years when no better coals were to be had, this seam was extensively worked both in Largoward and Lathallan, drained by day level.

The 4th or *Little Splint* was of fine quality but too thin to be workable, and the 5th or *Whitemyre Coal* was reported to be a splint coal with a white ash, but very little of it was worked although it was a level-free, dry coal.

The 6th or *Main (Largoward) Splint Coal* was an excellent hardy coal, which sold for 18s 0d per ton and survived being carted great distances, even to places where railway-borne coal was available. It experienced spontaneous combustion and gave off highly corrosive acidic waters that attacked any ironwork, and as the working matured the waters increasingly polluted the effluent streams. The last fitting in Largoward was lost due to the combination of fires and acid waters. A Parrot (not a gas coal) entered it next to the roof. In the Whitmyre field, there was 0.15–0.2m of this Parrot next to the roof. However, to the northwest, in the Falfield district, there was only 0.6m of Splint and 0.91m of the Poor Parrot coal and rums, of which little was put out and selling for 5s 0d a ton, while the Splint brought in 7s 6d. The Parrot was thickest where the day level workings ended, so that the Parrot mine in Falfield was less valuable than at either Largoward or Lathallan.

The 7th or *Black Coal* was a good house coal when fresh, selling at 2s 0d per ton less than the Splint, which was greatly preferred by customers. It was worked in the Whitemyre, Louisehall and Falfield pits.

The 8th seam or *Pilkem Coal* was too thin to be workable, but it was present beneath all of the Largoward coalfields.

Agreements were reached between the Earl of Lindsay and Mr Robert Ure that from Martinmas of 1899 work was to be carried out on the coals expected to be present below several parts of the Gilston estate. It turned out that none was worth fitting and working except the *Largoward Main Splint***.** If a regular field of it could have been found without its Parrot, it would have readily found a tenant. The gas coal at Gilston Mains was worth searching for, as it would have paid very well if the anticipated price of 16s 0d per ton could have been charged for it on the hill, as the mine was within a short carting distance of Largoward goods railway station. However, the gas coal was not easily found *in situ* despite many unsuccessful trials being made to find an undisturbed seam. The Baldastard workers found it repeatedly in the typical very disturbed condition of the luncarts, in irregular shapes and orientations.

The workings on the east of the Largoward Coalfield ended on a fault throwing the strata 90m up to the northeast, and the Whitemyre waste was worked out initially by a day level and later by steam engines leading to the workings of the Durhams and others at 90m depth. Unfortunately, no plans of the workings were made, but it is known that the area was very troubled with many faults, dolerite dykes and burnt coal. As far as could be determined, all of the coals beneath the Mains of Gilston March had been burnt in place.

The Whitemyre engine pit (N56.2624, W2.8604), which was 55 fathoms (100m) deep, lay to the northeast of the Cassingray Fault. To the west of the fault, Brown's No. 1 Pit met the Marl Coal at 9m and the *Largoward Splint* at 96m, while the Brown's No. 2 Pit reached the underlying *Black Coal* at 87m. The coals shallowed westwards with the seams becoming increasingly foul and poorer in quality, becoming unsaleable after 29 chains (654m) from the Gilston March. A series of boreholes in the Cordies Mealling area revealed that west of the crossroads in Largoward, the sedimentary rocks between the coal seams commonly contain traces of volcanic ash. The only coal seam shown on the BGS map to be outcropping between the Cordies Mealling and Largoward faults to the south of Falfield is the *Largoward Thick Coal,* which is younger than the *Marl Coal* at the base of the Limestone Coal Formation.

At the base of the succession were Thomas Brown's No. 1 Pit, situated 120m west of the church and 450m northeast of the crossroads in the village and Brown's No. 2 Pit, which lay to the southwest about 100m closer to the crossroads (Fig. 12.1). The principal seams worked from 1881 onwards were the Largoward Splint Coal in the former and the Largoward Black Coal in the latter. The top of the section of the Splint was 0.25m of rums followed by 0.05m of Parrot Coal above the 1.02m *Thick Splint,* which rested on a hard fireclay pavement. The coals were worked by room and stoop methods on the dipping bedding plane of the seam, inclined at gradients of 1 in 3 to 1 in 4 towards the northeast. The map with collation of detail by Forbes (1955) shows a small sketch of the area mined, but it is evident that this would have originally extended for some considerable distance to the northwest between the bounding faults. The transition from room and stoop to longwall working is shown to have taken place no more than 150m northwest of the No. 1 Pit base from where it was worked almost as far as the B941 road to Cupar. When worked to the northwest, the coals to the north became foul and the presence of wastes from old workings to the south forced the active coal workings to become progressively narrower, to less than 60m before a narrow north–south dyke of igneous material cut through the coals. However, the familiar troubles from baking, faulting and folding of the successions were less problematic in this considerable area.

A report in the East Fife Record of the end of March 1911 indicated that during some blasting and hand-working in Brown's No. 1 Pit an unexpected connection was opened up between the workings and the old mine wastes, which released waters into the working areas. According to Terris (undated) the waters rose rapidly and the pit was abandoned, fortunately with no loss of life. It had been believed that when the waters reached the foot of the main shaft, the main pump would be able to deal with the increased flows, and the

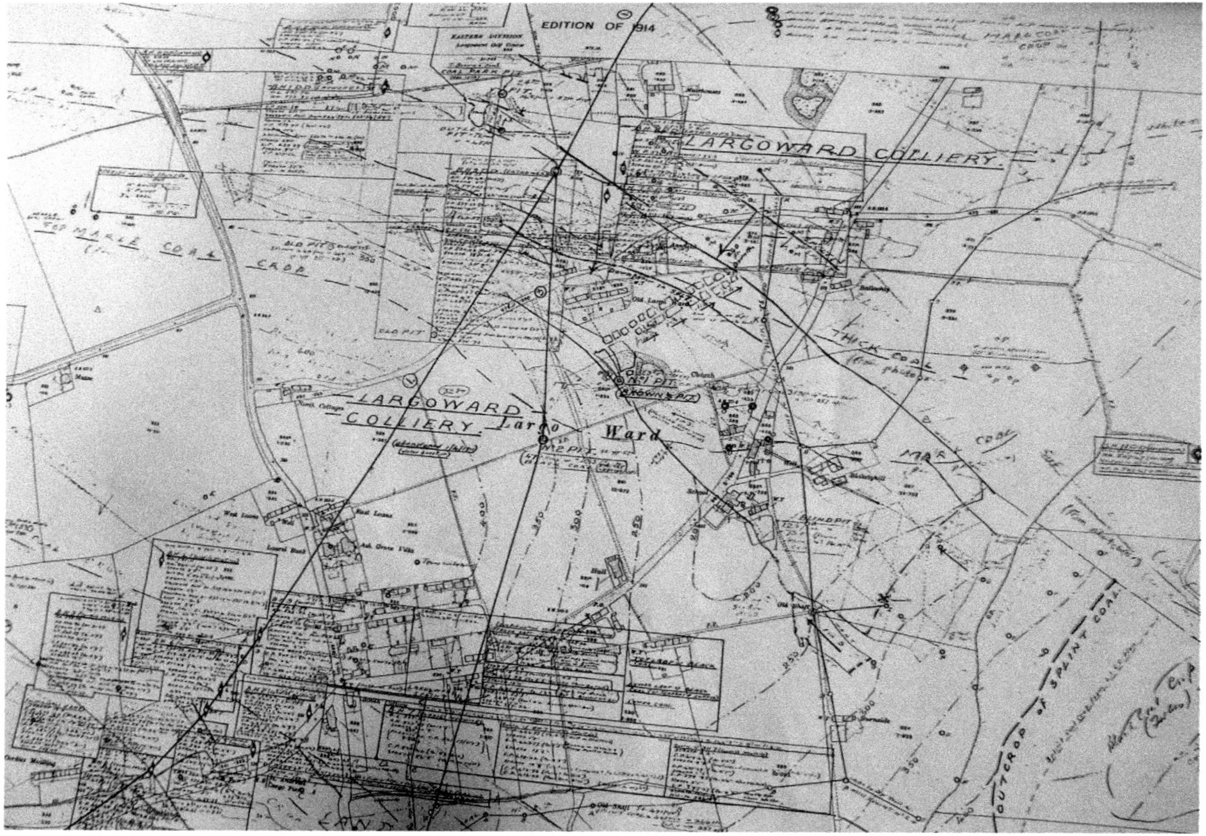

Figure 12.1 Locations of Brown's No. 1 and No. 2 pits at Largoward, from Forbes's collation map, 1955.

chemistry of the waters was expected to be benign. However, this was a forlorn hope and the waters were very aggressive, attacking all metals including the pump, which was lifted to the surface. The mine was closed and not thereafter re-opened.

The numbers of workers employed at Thomas Brown's Pits at various times during their working life are given in Table 16 below. Whatever the numbers engaged in the final stages, their being displaced and without work would have caused considerable problems for employment within the village.

Table 16: The numbers of colliers employed at Brown's Pits between 1881 and 1904.

Year	Underground	Surface
1881	81	19
1902	27	12
1904	38	11

Landale concluded that there was very little hope of any common coal being found workable on the Gilston estate, the only chance being of the Parrot Coal at Gilston Mains turning out to be a gas coal of value.

Falfield

To the north and west of the Largoward and Gilston properties, the grounds of Falfield (N56.2684, W2.8936) had a complex history of working known from at least the sixteenth century, when Falfield coals were supplied to the Royal Palace at Falkland and later to Cromwellian troops stationed at Struthers. The workings at Falfield lasted throughout most of the nineteenth century, as documented by reports of Landale dating from 1836 referring to 'old workings', and continued into the 1890s. The Falfield mines principally concentrated on the coals of the Lower Limestone Formation, passing up to the basal *Marl Coal* of the Limestone Coal Formation. The *Largoward Parrot* and *Black Coals* were divided into two units by the Cassingray Fault as its direction turned from southeast to northwest in the southeast, to west–east in the north. The faulting gave rise to the development of two coalfields, one to the north where the beds dip northwards at gradients of about 1 in 6, and the other to the south where the succession is folded into a shallow basin whose southwest to northeast-trending axis is partly outlined on the BGS maps by the Marl Coal. The coal-bearing sequence in both coalfields was essentially the same as that quoted above for the Largoward area (see Table 15, p.145) with the exception that the Marl Coal between (2) the *Thick Largoward Splint Coal* and (4) the *Thin Splint Coal* was not reported from Falfield, whereas the coals elsewhere in the broader district have normally been greatly disturbed and even burnt below ground. There was much successful working in the unaltered coals beneath the Dunnicher Law dolerite sheet west of South Bowhill Farm.

Landale recorded that in 1876 the North Field contained many shallow surface workings in the form of abandoned pits, in-fallen holes and signs of collapsed workings. The Field was initially drained to a depth of 31m by a Windmill Pit and later by an Engine Pit, giving a total drying penetration depth to 77m, completing the access to the whole of the Parrot Coal in this part of the field. Landale (1876) suggested that there might be much Black Coal still in place below this level. The presence of intrusive igneous rocks and the absence of coals detected in a series of four boreholes put down by the March at Braeside in 1874 was interpreted as indicating the presence of the customary troubles towards the eastern end of the field.

In view of reports from David Forgan, the former manager of Largoward, in 1836 telling of extensive workings of the coals beneath the protection of the dolerite sheet of Dunicher Law, Landale (1876) concluded that the hidden coalfield had probably been drained by the day level, carrying the waters to depths of between 56m and 70m. The coals would have been taken from that area by working dooks following down the dip of the coals so that there would be little remaining Parrot Coal. There would probably still be Black Coal at greater depth, but its quality was unlikely to have justified the fitting of additional workings at such a depth.

Landale continued to give advice on the Falfield coals with almost annual reports until 1890. He maintained a cautious approach, rightly not seeking to lead owners or tenants to unnecessary expenditure but always being clear in his recommendations. It is a little frustrating to have a substantial amount of information on the Falfield workings but to be unable confidently to relate the details to the ground concerned. Throughout, Landale refers to an overall map of the property on which he had evidently inserted a series of letters normally lacking place names or latitude–longitude indications. Unfortunately, neither Forbes (1955) nor the present writer have been able to locate that map, which would have held great interest. Attempts to locate it within the holdings of the National Archives of Scotland in Edinburgh have also failed. With that said, there are interesting figures available relating to the sizes of the workforce and the annual productivity of the workings through time, and these reflect the evolution of the coalfield and the likely impact of the eventual closures.

We know, for example, that once the Durhams of Largo had completed their extensive workings in the North Field they sold the coals to the Laird of Falfield. When operations were transferred to the South Field in 1877, following the drilling of exploratory boreholes, the main Engine Pit was established and fitted out at a cost estimated at £5000. A total of 15 colliers and 1 oncost worker were employed as the mine became operational. With this development came complaints from owners of riparian rights on local streams, principally

Mr Parsons of Kinaldie, claiming that the quality of the water in his stream would suffer, (apparently it did not). In the first year a total of 1422 tons of coal were recovered from the Splint in the *Whitemyre Coal*. In 1878, when difficulty was faced in selling the Main Coal (50% parrot and 50% common coal) it was estimated that in order to cover their costs the tenants would need to be selling as much as 10,000 tons annually, which was much more than the anticipated production.

In the new workings the pavement of the Largoward Splint in the South Field Engine Pit was 84m from the surface and the cross-section of the seam is listed in Table 17.

Table 17: Section through the Largoward Splint wCoal in the South Falfield Engine Pit.

Roof	Blaes	1.0m
	Pelt	0.5m
	Parrot Coal	7m
	Splint and Rough Coal	0.7m
Pavement		

The separated Splint was originally offered for sale at 9s 11d per ton on the hill, but even when the price was reduced to 5s 0d per ton, the stock built up due to lack of sales. The ordinary coal was also sold at a lowered price, but the carters still passed the workings to take the coals from Largoward or the Drumhead Mine at Peat Inn. By May of 1879 the workforce had grown to 18 colliers in the Largoward Splint Seam and 11 in the overlying *Whitemyre Seam,* and these workers were readily able to satisfy local demand. The uppermost coal, the Marl Coal, became accessible from the Engine Pit and the initial ingaun'ees were abandoned by the middle of 1879 when all outcrop workings ceased.

A second shaft was created during the summer of 1879 to serve as an air pit with a narrow shaft 1.4m by 1.5m in width. This shaft passed into wet faults with vertical throws of between 3m and 3.6m and linkages to workings from the first shaft were established using stone mines. The operations approached a substantial fault and the intervening beds steepened to near vertical, and there were signs that the faces were nearing igneous intrusions as the quality of the coals fell. Landale expressed disappointment in view of the fact that the fittings of the mine were in excellent condition, claiming them to be second to none in the country.

By May of 1880, Landale was reporting that although the Parrot Coal was available for working, it had proved a failure on the grounds that none of larger gasworks was prepared to buy it at an economic price. Very little had been sold for 4s 0d per ton, using the railway connection but leaving little benefit to the mine. The pumps had been withdrawn. Few of the 15 colliers working the seam before closure were retained to attempt to establish links to older wastes and on to the nearby No. 2 (Parliamentary) Pit. These connections gave access to the older drainage system and allowed the Marl Coal to be worked level free. The significant seams recognized in this mine were the *Marl Coal* at 47m; *Little Splint* at 65m; Upper Kinniny Limestone at 118m; *Main Splint* at 145m and the *Black Coal* at 168m.

Following the closure of the lower seam, 12 colliers began working towards the axis of the syncline in the rocks. However, before they could reach that point this coal, in turn, became steeply inclined and increasingly faulted and the coal quality fell still further. Another thin seam, the high-quality *Little Splint* 12m above the *Whitemyre Coal*, was shown to be an excellent house coal. However, at 0.4m thickness, it was only marginally thick enough to permit working, and having repeatedly suffered setbacks, the lessees were reluctant to continue.

Landale suggested that a bore put down close to the eastern Falfield March indicated the presence of a succession drained by a day level down to what he believed to be the Two Foot Limestone below the Whitemyre Coal and above the Main Coal. In 1882 he reported that the 1.8m-thick Main Coal had been successfully located at a depth of 50m in a new borehole. Although the borehole was remote from the Falfield Engine Pit, the winding engine previously used would be able to handle some coals from the nearby, but smaller, coalfield previously worked by Mr Durham. The detailed succession of coals in the main Largoward Splint in the rest of Falfield is given in Table 18 below.

Table 18: The succession of coals in the Largoward Splint Seam at Falfield.

Roof	Blaes	
	Pelt stone roof	0.05m
	Coal	0.20m
	Rums	0.46m
	Parrot coal	0.46m
	Splint coal	0.46m
	Rough coal	0.22m
	Coarse coal	0.15m
	Sklut holing	0.08m
Rock Pavement		

The succession leading into Largoward is largely free from even small faults, and 24 colliers worked 11 longwall faces. The Splint was originally offered at 8s 4d per ton as at Largoward. When reduced by 10% to 7s 6d a ton, the carts still passed Falfield to collect the identical coal at the higher price at Largoward itself. No further developments occurred until an Escape Pit opened.

By May of 1883 an air pit had been constructed to the coal at a depth of 36m and a total of 26 colliers and 3 oncost workers were exploiting the Largoward Splint in 7 longwall faces, 4 in two walls going east and 4 in two other walls revealing foul rock beside a dolerite dyke which had destroyed the coal. Until the foul coals were encountered, the enterprise had been doing well, having sold 5194 tons during the first half of 1883. Some of the foul coals were sold at 3s 6d per ton for lime burning, but this source of coal would become unusable in the near future. There remained five very good longwall faces with 0.50m of Parrot and 0.76m of Splint and other coal.

A year later Landale noted that the carts still passed Falfield to buy the same coals at Largoward, where the price was more than 20% dearer, but with the anticipated imminent closure of the Largoward, Lathallan and Peat Inn workings later in the year, he surmised that the situation of supply and demand might improve substantially in favour of Falfield. A new seam approximately 200m in length was to be opened in Lathallan, but the fitting would mean a delay of about two years before that would be fully operational, during which time Falfield would be the only producer of Splint in the district. During the fitting period at Lathallan there would be some production of the Marl Coal, whose quality would not rival the Splint Coal for sales.

By mid-1884 the two westward-going Falfield faces had run into foul coal and a fault with a 3.6m downthrow to the west. More foul coal was encountered by faces working towards the east, and the numbers of colliers in this part of the property fell to four working on three faces. A downward extension to the mine was proposed with a view to opening up the underlying Black Coal.

At the pit bottom, longwall working was established with three men working the faces in Falfield and four in Largoward to the east. The coal was a soft Cherry Coal 0.81m thick, which burned well but weathered quickly and sold for 1s 6d per ton less than the preferred Splint.

From the Black Coal a cross-cut stone mine was driven to the north to meet the Splint Coal and an air connection was opened up from the pit bottom. The seam contained 0.74m of Splint and 0.46m of Parrot, of which only the Splint Coal sold well. In the face of minimal sales of the Parrot at 5s 0d per ton, much of it was left in the ground. Overall, the seam worked near the eastern March Wall employed 25 colliers and 4 oncost workers. From the initial room and stoop workings, twelve longwall faces were opened up in fair coals to the west and four more to the east in poorer coals. By mid-1885, a total of 42 colliers were working the Splint Coal.

Exploratory boreholes put down to the east ahead of the rising workings in the upper coal 64m above the Splint found the Largoward Splint partly burnt. To the west the good quality coals encountered a fault with a 3.7m westerly downthrow, beyond which the coals had already been worked out and much of the area was under water. Towards the end of 1884 the extraction of material from the Black Coal seam came to an end when the seam caught fire and had to be stopped up. Five colliers remained working the Largoward Splint, three of them in an area of foul coal suitable only for firing the engine furnaces. The other two colliers were completing the final removals from the last patch of good coal.

The large new fitting was completed on Lathallan in mid-1886 to a depth of 113m, providing access to 'flat-lying' high quality *Splint Coals,* which were produced at a rate of 50 tons per day. In the west level the seam consisted of 0.46m of Parrot and 0.69m of Splint. In 1885–1886 sales of the Splint Coal were 9.91 tons at 7s 6d per ton, 1627 tons at 7s 1d per ton, and the Parrot Coal, 6,483 tons at 3s 0d to 5s 0d per ton.

The total of over 18,000 tons for the year of working was the greatest annual return from the mine and would not be exceeded, as the Lathallan fitting was then exhausted. At that time there were 51 colliers and 6 oncost workers in the Falfield Pit.

Increasing numbers of small faults were encountered towards the west and south, the largest showing throws of 3.7m, allowing most of the coals to be traced across the faults. To the east the layering curved gradually northwards to form an open fold. Here no fewer than10 longwall faces were worked in good clean coal, but when traced northwards the Parrot section fell in quality, producing sparks when burned. Eventually these workings were cut off by the same fault that terminated the Largoward Splint in the Cumberland Park area (850m northwest of the Largoward crossroads) beyond which the coals were foul and useful only for steam coal.

In his report of 1889, Landale indicated that the previous year had been very satisfactory with total production of slightly less than 10,000 tons, mainly of Splint and Parrot. However, all the coals from the bottom of the cross-cut mine had been extracted with the exception of a small area of Largoward Splint coal to the west of the March wall south of Cumberland Park, from which 1290 tons of coal had been produced in the preceding year. The workings here were stopped for fear of damaging the existing farm buildings.

In a further development to the west side of the pit, the coarse coals generally dipped westward at around 1 in 7 but the dips increased to 1 in 5 towards the substantial Peattieshill to South Falfield northwest to southeast-trending fault, beyond which wet sedimentary rocks were present. On the west side of the main dook 8 longwall faces were established and a further 7 to the east. The coals were 'indifferent' and were starting to accumulate on the hill due to lack of sales. Landale indicated that the rate of output would not be sustained, as six of the wall faces to the west had encountered foul coal. These faces had been duly sealed off to prevent ingress of water from elsewhere. To the east the Falfield Fault converged on the faces at a shallow angle and progressively narrowed the availability of the coals, which had assumed near-vertical layering as the fault plane was approached. The six final faces showed the coals becoming increasingly foul towards the fault.

Despite the rapid decline in the availability of the coals, the production during the final year ending Whitsun 1890 reached a total of over 10,000 tons. Landale warned that nevertheless the prospects for continuation of the workings in the Falfield controlled area were not good, and the lessees opted to break their lease at Martinmas (September) 1889, the earliest date on which they were allowed to withdraw.

Small workings to the northwest from Lathallan

Several largely independent workings northwest of Lathallan are believed to have started between 1860 and 1870, and the period of searching for more possible economic coals continued with many exploratory boreholes drilled in the 1880s in three areas:

1. An elliptical outcrop pattern 500m west of the Cordies Mealling Fault and similar distances north of Lathallan House and east of Cairn, which suggested the possible presence of a small basin.
2. The Appleton basin to the west of the Teasses fault and occupying the ground between Woodside and Baldastard.
3. The Bonnyton basin (N56.2536, W2.9538) between the Branxton and Teasses Faults situated between Bonnyton and Baldastard.

Coal extraction from these three areas continued after the closure of the Falfield colliery to the north and east. Logs of the boreholes are recorded on the BGS coalfield map, but warning was given by Forsyth and Chisholm (1977) that the depths and thicknesses quoted needed to be viewed with caution as for some, no recognized geologist had inspected the samples recovered to confirm their identities.

Cairn (N56.2496, W2.8827)

The elliptical outcrop pattern of the *Largoward Thick Coal* on the BGS 1:10,000 scale map (Sheet 40NE) serves to draw attention to the possible presence of a structural basin in the area. However, comparison of the land surface contours with the contours on the dipping seam reveals that the seam is not actually folded here, but rather that the outcrop pattern reflects the intersection of a south-eastward dipping planar coal layer with an undulating land surface. From the crest of the hill the land surface slopes south-eastwards at a lower angle than the coal seam. The ground steepens sharply near the centre of the slope before resuming its lower angle towards the bottom of the hill. This permits the coal seam to appear on the north-western side of the former railway cuttings and re-enter the ground to the southeast of the railway. The borehole logs indicate that the dip of the beds, including the coals, is uniform, dipping to the southeast towards the volcanic neck beside Lathallan Mill, as are the coals in the succession below ground.

The logs suggest that the Largoward Thick Coal becomes thinner towards the southeast, possibly decreasing from 1.12m to 0.61m in the 1km distance of separation between boreholes, whereas the older coals at greater depths maintain their thicknesses along the line from the outcrop to the Mill vent margin. At Lathallan Bore at No. 1 Pit of the Cairn Coalfield four coals were identified: the Largoward Thick at 7.6m; the Marl at 46.6m; the Largoward Splint at 145.8m and the Largoward Black at 168m, providing ready access to valuable reserves. The Cairn Pit was beside the railway line no more than 500m from the elliptical outcrop pattern shown on the map.

It is an interesting reflection of the times that a letter from as many as 26 colliers was submitted to Bethune George Walker Morison, the then Laird of Falfield, seeking permission to walk through a carefully listed series of fields on their way to and from work to reduce significantly the distance they would require to walk to and from their work if they were forced to restrict themselves to using public roads. These were workers living in Radernie, Lawhead and Peat Inn area needing to get to work at the Cairn Pit on foot in the early hours of the morning and to return late in the afternoon, very sensibly seeking to reduce their daily route of up to 10km by 20–25%. The language used and the entire approach is of great interest, as it is most unlikely that, today, quite such obsequious and over-courteous approaches of this sort would be made with such care. The metaphorical tugging of the forelock was almost certainly advised and possibly actually written by a legal adviser. Unfortunately, we do not know the outcome of the request.

Letter from 26 colliers seeking permission to walk across estate fields between Peat Inn, Radernie and the Cairn Mines while going to and from work.

> To: Bethune George Walker Morison Esquire of Falfield, Cupar
>
> Cairn Colliery, Largoward
>
> January 1893
>
> We the following parties being desirous of getting to and from our work at the coalfield of Cairn and to or from our several residencies at Peat Inn and Lawhead and Radernie by a nearer way than by the Public Road do hereby respectfully take the liberty of asking you to grant us permission during your pleasure to walk to and from our work along the fields on the estate belonging to you as follows viz. to such of us as reside at the Peat Inn along the road to Falfield Bank on

south side of Wellfield Park, West Meadow Park and Reservoir Park and to such of us as reside at Radernie or Lawhead through the middle of the field called 'James Dubb' field and along the top or south end of the field called East Frost Leas (on Bowhill Farm) and along the said road to Falfield bank and by the said eastern fence of said fields called Quarry Park, West Meadow Park and Reservoir Park. Upon your granting us the said liberty we hereby undertake not to trespass through any other part of your property nor injure the same, and we hereby agree both individually and collectively to stop from walking across your fields upon your asking us or any of us to do so, and that without any form of legal warning or notice of removal or other process to that effect.

Alan Small, Radernie; Andrew Donaldson, Radernie; William Balfour, Peat Inn; George Balfour, Peat Inn; Alex Brown, Peat Inn; James Robertson, Radernie; Alex Ireland; Thomas Duff; James Pryde Jnr; John Ireland; Walter Wilson; James Pryde SNR; George Pryde; John Pratt; John Robertson; John Robertson, Westfield; David Walker Kineilson; David Donaldson; George Donaldson; John Pryde; George Berwick; Thomas Donaldson; George Suttie; Peter Webster; George Drylie.

John Pryde, Witness.

The Appleton Basin

The second nearly circular pattern is a genuine geological structure, the Appleton Basin. The basin structure has been preserved or formed as a result of its occurrence between a series of faults. To the south, the east–west fracture of the Branxton Fault between Branxton Farm and the weir on the Gilston Burn some 2.5km to the east throws the Upper Limestone Formation down towards the south, where volcanic agglomerates associated with the Largo Law centre are to be found. The displacement is largely duplicated by the sub-parallel Teasses fault, which crosses the area between Baldastard and Appleton. To the east of the basin a north–south trending fault with a downthrow to the west outcrops along a line between Woodside and Gils Law, and this permits the floor of the basin, complete with the Lower Limestone Formation, to reappear to the west. At the centre of the basin the youngest part of the succession is a thin representative of the lowermost part of the Limestone Coal Formation including the youngest coal, the *Appleton Splint*. Worked out many years previously, Forsyth and Chisholm (1977) reported it to have been up to 1.20m in thickness. This coal is probably equivalent to the Marl Coal, which has been widely worked in the Largoward area. About 15m below the Appleton Splint Coal is the *Appleton Ell Coal,* which is shown by borehole logs to have been 0.5–0.6m in thickness. The term 'ell' is an old Scots measure of length, a form of yardstick, approximating to 0.94m. A problem of identifying the individual coal seams arises here. Thus far the lowest seam encountered in this part of the whole area is believed to be the *Largoward Black Coal,* which is overlain sequentially by the *Largoward Splint Coal,* the Mid-Kinniny Limestone, The *Appleton Ell Coal* and the *Appleton Splint Coal.* The Mid-Kinniny Limestone has not been detected on the south side of the Teasses Fault to the west of the former railway line, possibly due to lack of exposures in this small area. North of the Teasses Fault the stratigraphic sequence west of the Appleton Basin shows the *Appleton Ell,* above the Mid-Kinniny Limestone, above the *Largoward Splint,* the *Largoward Black* and the *Teasses Main Coal.* The *Buffo Coal* of Teasses referred to by Landale (1879) is in a similar stratigraphic position to the coals of Radernie, above the St Monance Brecciated Limestone, but little detail is available on the deposits in the Teasses area.

The Bonnyton Basin

Northwest of the basin centre, the geological maps show that the layers are all arranged dipping south-eastwards at about 15°. As a result, progressively older beds outcrop towards the northwest in a sequence from the Mid-Kinniny Limestone, the *Largoward Splint* and the *Largoward Black Coals.* The Charlestown Main Limestone, which is present in the succession at the coast, is not known either at outcrop or in boreholes here. The 1.12m thick *Teasses Main Coal,* a probable equivalent of the upper coals of the Radernie succession, outcrops north-eastwards from 150m north of Teuchat's Toll (N56.2552, W2.9614) cross-roads and may be traced towards and beyond Hill Teasses House. To the northwest the Teasses Main Coal, the St Monance Brecciated Limestone and the upper part of the Pathhead Formation are progressively overlain by the Teasses Common quartz-dolerite sill. The reappearance of the St Monance Brecciated Limestone and the underlying Pathhead Formation in the Fleecefaulds–Bankfoot area north of the sill suggests that some form of faulting is present within or below the sill mass to permit the repetition.

The stratigraphical problems of this third coal-bearing basin centred at (N56.2536, W2.9538) have been partly addressed above. Again the structure is preserved between both the Branxton and Teasses Faults, but smaller northwest to southeast-trending fractures are present in the west. The principal coal seam in this basin is the *Largoward Splint Coal,* termed the *Blue Eye* in Largoward, which is partly repeated by the faulting, and the *Teasses Main Coal* is also shown to outcrop. Presumably the Largoward Black Coal is present, but the faulting does not permit it to appear at the surface.

Several additional thin coals outcropped above those already addressed, but they were mainly towards the eastern end of the plantation. Landale (1879) commented that all had been worked from shallow pits that had been abandoned at least half a century previously, towards the start of the nineteenth century.

Teasses Common

The lower seams of the south-eastward dipping coals described above from the Baldastard and Bonnyton areas continue northwards beneath Teasses Common. The upper seams of the Lower Limestone Formation slope away from the common and consequently are not represented, so it is principally the lowest parts of the succession that abut or underlie the major quartz-dolerite sill that forms the surface material of most of the common. The uppermost layers seen include both the *Blue Eye* seam and the *Largoward Black Coal* below it. Both of these are above the *Teasses Main* and *Teasses Under Coals.* They were believed to have been worked down as far as the standing water level because the price for sales was not sufficient to make it worthwhile to install pumps to remove the waters. Mr Martin (tenant at the nearby lime works) was then using some of the coals for lime burning.

A series of small-scale maps compiled into the very useful reference work by Michael Martin *et al.* (2007) show the former presence of many working coal pits and a range of shafts in the Bonnyton, Baldastard Mains, Teuchat's Toll area. Most of these workings are thought to have started operating about the beginning of the nineteenth century and similarly, most are believed to have closed by the middle of that century. The outcrop of the 1.47m thick Teasses Main Coal follows a northeast to southwest line for 2km from south of Hall Teasses towards Teuchat's Toll, where it becomes involved in a series of small faults. Towards the north, the coal converges on the margin of the Teasses Common sill, and as a result becomes increasingly blind and foul as the bitumen has been largely driven out. Where the seam passes beneath the sill, less alteration would be expected as the heating from the intrusive igneous masses preferentially rose into the rocks above, but there is no directly reported observation of this phenomenon locally.

To the north of the east–west trending part of the Cassingray Fault, there is no evidence of the presence of the sill. Instead there is a broad area of land leading to Newbigging of Craighall with its now abandoned quarries in the St Monance Brecciated Limestone, the outcrop of at least one of the *Teasses Coals,* and the Gathercauld intrusive masses to the east. At the foot of the columnar jointed sill are outcrops of the Mid-Kinniny Limestone and of the *Largoward Splint Coal,* marking the site of the final coalfield to be addressed.

The Newbigging of Craighall Coalfield

The majority of the coalfields that have been examined above have been on a relatively large scale for which the detailed considerations of the field development could not have been addressed within the limited scope of this text. A set of papers, largely derived from Landale's work, addresses a short-lived coalfield located in a small fault-bound depression in the prominent dolerite sill of Gathercauld Craigs to the east of Newbigging of Craighall 3km south of Ceres. These papers provide an insight into the prior assessments of the area required before exploitation of the coals could begin, and illustrate a range of not unusual physical problems encountered as the works progressed between 1863 and 1903. The term 'bigging' is a commonly used old word for 'building', so Newbigging of Craighall was a new building on the Craighall estate in the east of Craighall Den, where the former estate castle, known from AD1410 and enlarged in 1647, was demolished in 1957.

The earliest mining in the north of the district is known to have extended from well before 1749, when reference to long-abandoned workings between Ceres and Callange was made in the Banfield Coal Book (1761). The earliest mining involved removal of the outcropping coals, no fewer than 17 seams of which were listed by Beaumont (1806). This developed the use of bell pits and shafts, leading to room and stoop operations with the introduction of cross-cut and multi-level mines near Callange itself. The workings at Callange and South Callange were all drained by a day level leading north, and which still discharges into the Kinninmonth Burn (McManus, 2010). Some of the mines in the north were worked on the grounds of the Craighall estate. Other thin coals were known and worked both in Craighall Den and also to the north of the Den, principally to produce coals for lime burning from the locally worked St Monance Brecciated Limestone (Dixon and Ball, 1808).

Knowledge of the limestone and coal at Newbigging of Craighall is derived principally from an account given by Landale (1837) who suggested that there was likely to be a small coalfield with at least two thin seams of Teasses coals lying in a workable position east of the small established limestone quarries on Newbigging Farm. He noted that the adjacent hill masses to the north and south were of igneous material, which might have led to blinding or breaking up of the coals. Throughout the Largoward district the proximity of intrusive rocks has been shown repeatedly to have led to the chemical and physical destruction of the coal seams. In this case, the thick sill of dolerite rests above the coal deposit and is separated from the coals by dipping, coal-free sedimentary rocks. Although the dolerite would have had the capacity to alter the coals, when it was intruded the heat from the sill would have risen to cause changes in overlying material; the laws of physics decree that heat rises, allowing the underlying coals to survive unscathed. Along their northern and southern boundaries, where faulted contacts permitted direct contact, baking would be anticipated.

All went quiet for almost 50 years until John Ritchie made preparations to open a new limestone quarry, writing to the factor, John Flockhart, offering to pay a basic rent of 1/- per ton above an output of 700 tons. When inspected by Landale and the estate owner, Col. Anstruther Thomson of Charleton, the quarry turned out to be a mine in the limestone, the layers of which dipped eastwards at a gradient of 1 in 5.

On 11 February 1895 a letter from Mr Flockhart to the Landale, Frew & Gemmell (L., F. & G.) Partnership indicated that some time previously Col. Thomson had let the limestone on Newbigging of Craighall to Mr James Simpson, a former tenant of the Teasses Limeworks, giving him permission to bore for coal in the two easternmost fields of the farm. Mr Simpson had then indicated that he had found a seam of good quality coal and provided samples of the material from the trial pit. He wished to work the coals and was seeking some financial assistance from the estate. Landale duly replied, recommending the putting down of a pattern of boreholes in order to assess the likely success of the enterprise before the work started (Fig. 12.2).

On 27 February Mr Flockhart notified Landale that the boreholes had located good-quality splint coals between 0.50m and 0.58m in thickness in six of the eight borings at depths of 5.5m, and the seventh showed coals at 11m. There was a strong 0.2m-thick pelt roof of strong, slightly bituminous shales overlain by 1.2m of strong blaes. Beneath the coal was a 0.6m-thick layer of black fireclay above a thin layer of coal. Mr Simpson sought security of £150 from Col. Thomson to enable him to open up the field. A full survey in March 1895 by L., F. & G. noted that the borehole pattern had demonstrated that the coals extended for at least 200m along the direction of strike, i.e. north to south.

The coal seam lay approximately 750m east of the limestone working and dipped at 1 in 5 eastwards. They confirmed that there had been no evidence of burning by the overlying igneous intrusions in any of the samples of the coal seen. The near surface of the material was firm and splint, and it was anticipated that a considerable percentage in the seam would yield round coal. Samples tested showed that although the material appeared to be almost a parrot coal, it burned quietly and yielded a large quantity of gas. Other pieces were of cherry coals that burned well, yielding 3.5% of ash, i.e. a clean household coal. The coal did not readily separate from the overlying hard pelt layer, but the fireclay below would be suitable to be holed beneath the coal. There was water from the surface, but hand-operated pumps had been used to keep the water level down in the trial pits.

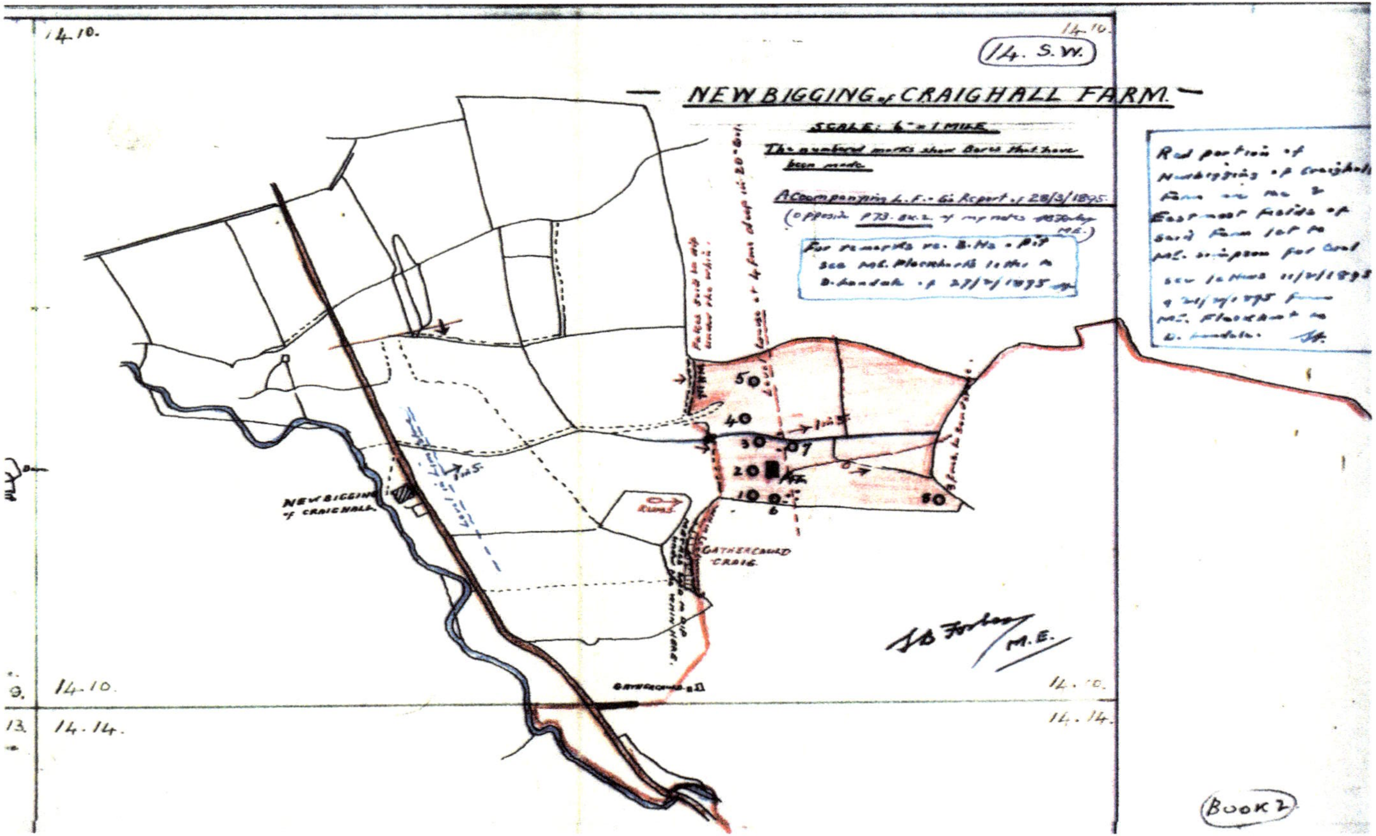

Figure 12.2 Early exploration map (1895) of the Newbigging of Craighall mine development area.

A second coal seam 0.77m in thickness and 10m above the first was found in two of the deeper bores. A third coal, 0.46m thick, thought to be the lower of the two *Teasses coals,* was located 37m below the limestone, but neither the *Lady Mary Coal* nor the younger Teasses coal were found. A layer of poor-quality 'rums' or foul parrot coal was noted below the coal outcrop ear-marked for development.

L., F. & G. believed that although the masses of igneous intrusions were close at hand, it was possible that the coals would extend unaltered beneath the Gathercauld sill, as had been the case at Baldutho further south. They recommended that the coal should be opened cautiously and suggested that a 22m deep shaft should be sunk to the east of the excavated ditch, which would give a working face of coal of 30m to the rise on each side of the 3.7m pit and give a field that should produce 24 tons of round (66%) and small (33%) coal per day equal to 6000 tons per year. Above the level of this pit the coals should last for three years at that rate of working, after which a deeper pit would need to be sunk. The cost of such an operation was estimated at £200.

Table 19: The initial estimated economics of the Newbigging of Craighall Coalfield

Expenditure		Income	
Production, hewing & drawing	3s 4d per ton	Sale price of round coal	9s per ton
Brushing roads & cutting	9d per ton	Sale price of small coal	5s per ton
Pit wood	2½d per ton	Average price	7s 8d per ton
Oncost	1s 0d per ton	Profit: 7/8d less 6/2d	1s6dper ton
Furnishing & repairs	3½d per ton	Profit on 6000 per year	£450
Interest & redemption	2d per ton		
Total cost	6s 2d per ton		

By the time of the next inspection in September 1895, the second pit, complete with horse gin and winding frame, had been constructed and, as foreseen, there was trouble with inflowing water. This was being removed by a small steam-driven pump placed on the surface. The 5.5m deep pit reached the upper coal where levels had been opened out to 20m on each side of the shaft without encountering faults or other problems. The coal was 0.74m thick, the uppermost 0.2m of which was a hard splint.

The shaft continued downwards, meeting a rums and foul parrot seam at 11m, suitable for little more than lime burning. From that level a horizontal cross-cut stone mine towards the dip direction had been driven to meet the upper seam 28m east of the shaft. Here the coal was harder and cleaner than nearer the surface. Levels were now to be driven to both the northern and southern flanks of the coalfield from point A on the map (Fig. 12.3) and linked

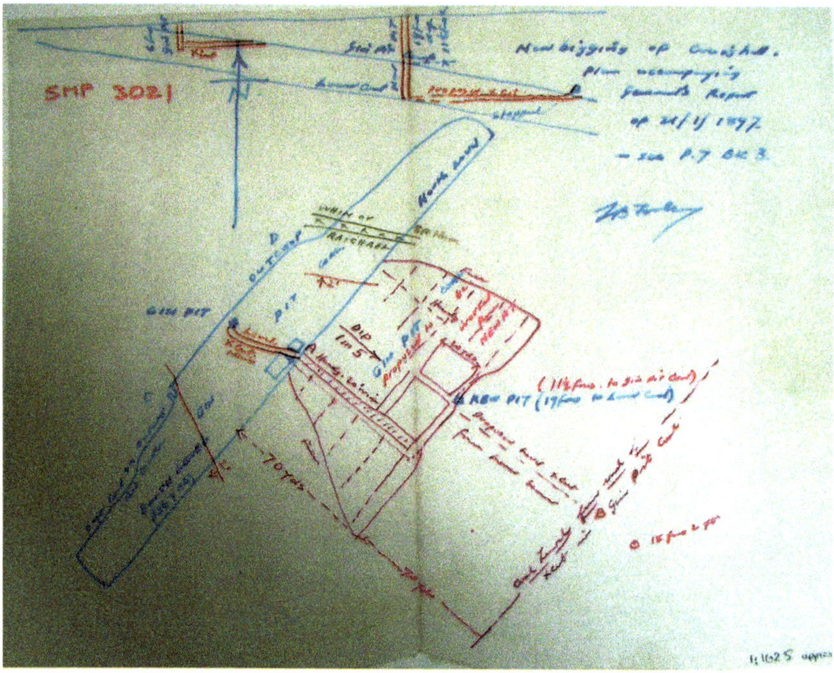

Figure 12.3 Late stage of development (1897) of the Newbigging of Craighall mine workings (Landale, Frew and Gemmell).

to the 64m-long down dip heading to connect with levels driven from the base of the New Pit shaft.

The exploratory boreholes demonstrated the dimension of the coalfield. Taken at surface level 10 acres (4ha) measured for both the upper and lower seams could produce 59,000 tons of coals and dross in a working period of nine years. It was recommended that the coalfield should be let for an annual rent of £60, with a royalty of 1/9th of the total value of the sales at the pit and 3d per ton for the small coals used in the Newbigging Limeworks. Working in the *Gin Pit Coal* (the upper of the two seams) was stopped in the northern level during November 1896 when the road roof became so low that the coal could not be drawn out of it.

The roof of the south level had also become low, but working continued despite it also becoming very wet. By the end of February 1897 the four colliers and two drawers working in this level had advanced 125m from point A. The coals had thickened to 0.92m, the upper 0.21m of which was of splint. Three other colliers were working in soft coal in two rooms only 2.1m wide near the outcrop at C. Unfortunately, this near-surface working permitted the entry of water into the workings in wet weather.

The New Pit sunk 90m to the dip from the Gin Pit gave access to a 64m-broad stretch of the Gin Pit Coal. This working was drained by a small steam engine and pump. The *Gin Pit Coal* was 0.8m thick at 21m and the shaft continued down to meet the *Lower Coal* at 35m. This coal was 0.5m in the original crop pit of Simpson but thinned to between 0.4m and 0.45m at this depth. The southern face of the Gin Pit was to be worked as soon as possible to avoid it also being flooded.

By 1 February 1898 only two rooms were working. That on the north side was stopped at a 0.76m upthrowing fault. Gemmell reported that the pumps in use were being pushed to their capacity, one discharging 450 litres per minute, substantially above design capacity. To enable two rooms to be worked satisfactorily would have required the introduction of larger pumps able to lift 570 litres per minute to heights of 18m.

Although the upper workings in the Gin Pit were abandoned on 1 February 1898, the lower level New Pit continued in productivity, but with increased pumping capacity, which also aided drainage of the open surface trenches excavated around the workings. The coal was accessed from openings off the shaft, which enabled the working of eight wall faces, two to the north and six to the south. Together they provided more coal than could be sold, and a temporary storage bing on the surface was noted. One underground oversman (deputy), two pit-headsmen and two engine men were employed in addition to an unspecified number of miners, probably at least one to each of the eight faces.

By 12 April 1899 sixteen miners (men and boys), an oversman and one bottomer below ground, two pit-headsmen and one engineer worked in the New Pit extending away from the shaft. In both the northern and southern workings small dislocations of the coals referred to as 'loups' were encountered where the seams thinned to 0.4m before opening up once more beyond the loup. On the south side the working had stopped at a fault with a 3.7m upthrow to the south, but this was being bypassed. By this time no less than 1 hectare of the available coals had been mined, and there remained about a further 0.5 ha of extractable coals to be worked up towards the Gin Pit level.

By 29 November 1899, the south face had been extended to 137m from the pit bottom. The north face had been worked out for 146m on the New Pit level. About 70m from the shaft base the seam had been cut through by a 2.5m-thick vertical 'Ratchel' dyke of a softer than normal dolerite, beyond which the coals had been burnt and rendered unworkable for 20m. For another 32m beyond the dyke, the coals were again workable until a large east–west fault, traceable up to the Gin Pit level, cut across the seam. On the south side about 50m of the seam breadth remained to be worked but several small (0.6m) faults had been met. To gain more workable resources the known 3.5m throw fault would need to be breached.

In general, the underground workings were considered well ordered, the roadways of a good height and ventilation good. Gemmell reported that 21 men and boys were working below ground and three on the surface.

By August 1900 on the south side of the mine a new level had given 7.5m of workable coals before meeting the 3.5m upthrown fault. This fault had been broken through, revealing a further 30m breadth of coal to be worked, but distressingly the seam was now heavily broken by small faults.

There was concern over the expenditure on wages in comparison with the work that had been carried out and the income generated from sales of the coals. Gemmell reviewed the activities of the miners, pointing out that the matter of payment was principally brought about by the fact that all of the colliers were paid at the same tonnage rate whether the coals were easy or difficult to work. Additional supplements were paid to those working in difficult places. In the three main levels all received extra payment for cutting the solid coals to the rise and for brushing out beyond the roof to give height to the roadways. For example, payment of 6s was made per 2m of roadway but the cost per ton varied with the thickness of the coal and the breadth of the wall faces being worked. For a thick coal and normal length of wall 8s 4d per ton was paid, but where the coal was thinner and the wall length shorter, the road advanced a greater distance for every ton. He also discussed the problems of assessing the work carried out and the wages paid individually to ensure that all was fair to the workers when working longwall faces. He concluded that the payments given at Newbigging of Craighall were at normal levels in the industry and that the only way to ensure total fairness to the workers would be to show precisely what tasks had been done during the period under consideration. This would in itself incur further expenditure through the need to employ someone specifically to undertake that assessment.

An inspection dated 7 April 1903 revealed that work was continuing in the 36m deep Ratchel Pit, but the numbers of staff then employed had fallen sharply to seven miners, one drawer, one boy at the pit bottom besides the oversman, an engine man and a pit-head man. On the south side all work had been stopped once the final faults had been crossed to reveal at least 4.5m of crystalline igneous rocks of the dolerite sill that formed the bulk of the Gathercauld Craigs mass.

Boreholes drilled nearby failed to locate any additional reserves, but a second series of holes did locate a coarse limestone believed to be about 9m above the *Teasses Main Coal*, accompanied by an under coal known to be 4.5m below the Main Coal elsewhere. The Teasses Main Coal in the vicinity was not believed to be worth working and it was recommended that the operators look to the north, towards Callange, where continuation of the coals already worked might be found beyond the west–east section of the Lathones Fault at Baldinnie.

The coalfield that lay approximately 220m east of Newbigging of Craighall Farm had been shown to consist of two seams extending north–south for approximately 270m and dipping eastwards at 12° between two substantial faults that defined the limits of the areas within which the coals had been preserved. The aim of this short concluding section on the coalfields has been to follow the discovery, exploration, evaluation, the history of the working of the resource, and the eventual closure of a small set of workings. Sadly, the enterprise lasted for little more than six years. Yes, there were other thin but not very productive coal seams nearby in Craighall Den and north towards Ceres, but none of these is known to have experienced even the brief success enjoyed by the Newbigging of Craighall site. The enclave of the Pathhead Formation sandstones stretching from Bandirran into Craighall Den signals the western margin of the Largoward Coal Basin of the East Neuk Coalfield.

Chapter 13

Closing thoughts

The East Neuk of Fife was blessed with a mineral resource that was relatively easy to access, initially from the coasts and valley sides. However, to the east of the line between Largo Law and Cupar, the coals were distributed in relatively thin seams, unlike those of Central and West Fife. We do not know when the coals were first worked systematically in the East Neuk, but they were used for working metals in the north of England during the first and second centuries AD during the height of the Roman occupation. At that time an advanced base and assembly area for the legions seeking to extend the boundary of the Empire towards the north was established at Carpow (*Horrea Classis*) beside the junction of the River Earn and the Tay Estuary. During preparations for excursions to the north some of the coals supplied from North Shields would have been used in the military workshops, but there is no record of the coals having been mined in Fife at that time. It is recorded that the Roman Empress Julia Domna entertained the troops at Horrea Classis, possibly using coal-fired heating systems but again, not derived from locally mined coals.

During the Dark Ages and the early medieval period there were some attempts to work the coals for heating the major stone-built buildings, the castles, churches (St Monans) and cathedrals (St Andrews). The issuing of royal charters to formalize the industry in the thirteenth century may well have been an early means of raising taxes beyond those raised under the feudal system. Most of the charters were issued to ecclesiastical settlements, sometimes permitting the coals to be taken for their own use and not re-sold at market (Dunfermline). These added restrictions serve to remind us that the coals were not worked in isolation. Whatever was happening in the nearby world exerted some influence on the mining process.

The coal-mining industry was starting to become established in the fourteenth century when a series of outbreaks of plague, said to have been spread by infestations of flea-ridden rats, affected most of Britain and the continent. The several waves of deadly infection led to destruction of 20–30% and more of the population. The feudal lords lost many of their workers, and in order to maintain food supplies many of the colliers were redeployed to the surface to work in the fields. The output from all mine-workings fell drastically and remained at a low level until the mid-sixteenth century, well after the plagues. As a consequence of flooding by groundwaters and repetitions of the start-stop form of the working, some of the earlier mines were lost, only to be discovered in later workings, often leading to problems for the new miners.

As the industry was recovering from the severe medical problems, religious fervours began to sweep the continent, and Fife was no exception. The reformers and traditionalists from two factions within one religion proceeded to fight and demanded that the estates should send their men to support one side or the other. As a result, the population of able-bodied men in Fife was again sharply reduced, the industry stuttered once again, and more mines were lost. Mercifully this was the last major disruption to the industrial development until the two world wars of the twentieth century. In place of disruption, the Industrial Revolution was starting, with greatly increased demand for coals to support the new industries. Throughout most of the eighteenth and nineteenth centuries, coal mining flourished and most of the reserves of coal in the East Neuk were worked out so that, by the start of the twentieth century, there had been a drastic reduction in productivity and the horrors of unemployment were hitting the population. Few of the mines survived into the 1920s and still fewer remained operational thereafter. The very small operation at Radernie continued until the Nationalization of the coal industry in 1947. The East Neuk area has now returned to largely agricultural working, and supporting commuter-based industries such as the University of St Andrews and others outside the East Neuk in Glenrothes and Dundee. To add to the problems of the area, with its greatly expanded student numbers and burgeoning tourist industry, the railway system around the coast, the line between Leuchars and St Andrews and also the Central Fife Railway have all been closed.

What of the future prospects? During much of the time that coal seams were being sought, there was considerable interest in the potential of shale oil, such as was worked in the Lothians. Many sites were examined and most found not to have sufficient oil content in the shales. The levels of oil known at some of these sites may merit a re-examination of this potential resource. In the light of modern technical advances, a re-assessment of the materials could well lead to encouragement for development of another indigenous oil-shale industry. The dreaded word '*fracking*' doubtless intervenes in any discussion of what has become a highly valuable industry in North America and elsewhere. In Britain, the issue of fracking has become fiercely resisted by concerned lobbyists. Would the fracturing of shales below ground lead to the development of earthquakes? What liquids would be injected into the rock layers? What additive chemicals would be injected with the waters? Are we certain that the injected liquids will not contaminate our groundwaters, and would they emerge into the Forth sea area?

Perhaps more to the point geologically in the East Neuk, we know that the coal- and shale-bearing rocks are concentrated into a relatively thin 450m thick upper skin, which is riddled with curving faults. However, we have scant knowledge of what lies beneath this thin layer. The surface few hundred metres of rock certainly contain small gas concentrations, and oils are known from several of the small folds present, but exploration by the oil industry towards the end of the twentieth century failed to locate any sufficiently substantial reserves to encourage onshore developments. Our knowledge of the rocks of the more than 1000m of the Strathclyde Formation suggests that what lies beneath the coal-bearing Carboniferous rocks is likely to be predominantly of mixed assemblages of rock types, of which few, if any, carried the hydrocarbon materials that formed coals. The underlying Devonian rocks preserve few deeply oxidized fossil remains, and the reducing environmental conditions needed to preserve organic remains are missing, so the potential source rocks needed to supply the natural gases appear to be absent.

Bearing in mind the considerable disruption of the rock successions seen in the East Neuk as a result of folding and fracturing of the sedimentary rocks within which the oils and gases may have developed in the past, and their subsequent intrusion by the vast mass of basic igneous material of the Midland Valley Sill, it is sensible to ask about the possible survival of the hydrocarbons when their host rocks and structures are invaded by very hot molten materials. In virtually all of the mines examined in this text, there is evidence of the cooking of the rocks in thermal metamorphism to the extent that many of the coals are seen today as having been hardened, and the associated shales greatly disturbed and turned foul to create the rums and luncarts as the gases were driven from the coal seams on the approach of the magmatic material. The intrusions did not all occur in one single event but as repeated incursions spread over a period of six or seven million years (Read *et al.*, 2002). So the time over which the processes of degradation of the deposits occurred was geologically lengthy, with at least three phases recognized in the East Neuk area.

The proposed attempt to release energy from underground burning of the coals beneath the bed of the Forth seeks to provide a second direction of attack, by burning the coals below ground to release the heat energy without removing them. The numbers of sites at which the technique could be used will be limited by the damage already inflicted upon the coals by thermal metamorphism, as indicated above. In many places it will be clear that Mother Nature got there first. At present not enough detail of relevant exploration has been made available by Cluff, but it does appear that there would be much less injection of liquids involved in the process of gasification of the coals below ground than involved with the fracking proposals. It will be necessary to watch carefully and weigh up the techniques – challenging if insufficient information is provided. Until a suitable level of information is presented, it would be wise not to adopt a fixed view for or against either of the proposals, and not to leave the comfortable fence. Until that day, we need to gather as much detail as can be obtained. At that stage, and not before, a clear and defensible position for or against may be adopted.

Glossary of mining terms used in East Fife

A

adit a horizontal or low-angle gallery or roadway.
anthracite a bright coal that is hard, burns with a pale flame and gives off intense heat. Has high carbon and low volatiles content.
anticline an arch-shaped structure in which the rocks dip away from a linear central axis.

B

back to throw coal back along the working face towards the head of the road.
bad roof a roof of cracked or soft material, which is susceptible to falling in.
band a thin layer of rock whether within or outside the coals.
bank surface of the ground at the pit head opening.
basin a depression or trough towards which rocks dip from two sides.
bastard an impure, often gritty, coal seam.
bearer the wife or daughter of a miner who carried coals up the shaft to the hill.
bed a layer or stratum of rock. A coal seam is a bed of coal.
biggin an accumulation of wastes from coal workings within worked room.
bigging building.
bing a heap of coal or coal wastes.
blackband a carbonaceous clay ironstone, self-calcined, often in nodules.
blaes dark-coloured bituminous mudstone or shale.
blind coal a coal heated by igneous intrusion, which loses volatiles and burns smoke-free.
bore to create a hole in the rock, usually by drilling or chiselling.
borehole a hole drilled into the rocks during exploration for mineral reserves. Also for release of drainage water or gas, and in shot firing.
brass iron pyrites in lumps or veins within the coal.
brasso/brassy coal with veins of iron pyrites.
brush remove part of the roof or pavement to heighten the roadway.
brusher worker employed to attend to the brushing, whose job is to brush.
bye pit a subsidiary or secondary pit.

C

calm pale-coloured blaes hardened and bleached by nearby igneous intrusion. Formerly used as early pencils.
cannel dull coal giving much gas and long-lasting bright flame.
chain length measure. Scots chain is 74 feet long, Imperial chain is 66 feet.
cherry coal resinous, shiny, cubical coal. Burns readily. Popular for domestic use.
cleat a natural fracture seen repeatedly in a coal seam.
collier a coal miner.
craw coal a poor, thin coal a short way above a major coal. Acts as an indicator.
cross-cut a tunnel driven across the rock layers to link between different seams.
culm very small-sized particles of dross coals.

D

daugh soft fire-clay associated with a coal seam in which holing is made.
day level a water level driven from the surface.
dip the angle at which rock layers are arranged from the horizontal.
dip-head level roadway within the body of the coal seam.
dook underground roadway that leads down the dipping rock layers.
drawer man or boy who takes coals from the face to the foot of the shaft.
drift a tunnel sloping downward from the land surface.
duff coal dust or dross.

E

exposure where a rock can be seen at the land surface.

F

fakes a laminated, banded siltstone or muddy sandstone.
fathom a measure of length. One fathom is six feet (1.83m) in length.
fault a fracture plane along which the rock succession is displaced up, down or laterally.
fireclay a refractory clay usually found beneath the floor of a coal seam.
foul coal coal that has been altered by proximity to igneous intrusion.
free coal coal that is readily broken and burns well.

G

gin a mechanism for raising coals to the surface using horses, water or wind.
goaf (goaves) waste material below ground from an area of mined coal seams.
graip or herp a sparred shovel used for filling coal.

H

head coal The layer of coal immediately below the roof of the seam.
herp shel/graip with bar blades for primary sorting of coal by size.
heugh shaft of a pit or mine.
hill the land surface at the top of a pit.
hitch usually used for a small fault with relatively minor displacement.
hole to cut below or above a coal seam to help dislodge coal from the face.

I

ingaun e'es open cavities in cliffs or hillsides marking entrances to working mines.

K

kingle a very hard rock. Usually sandstone with silica or carbonate cement.

L

leaf a layer of coal within a coal seam.
level a horizontal underground roadway along the strike of the rocks leading past galleries or rooms worked for coal. Gives ventilation or haulage.
level free a roadway within the upper part of the coal seam above standing water level.
load an early measure of weight for coal. Varied in weight by locality.
longwall Mining method in which the entire resource is excavated.
luncart large-scale structureless mass of disturbed bituminous shales with coal fragments.

M

marl a calcareous clay or clay-rich limestone.

march dyke a mound marking the boundary between two adjacent estates.
mine a sloping or vertical tunnel leading to coal seams or metal ores.
mine mouth the site at which a mine goes below ground.

O

oncost labour costs for working the mine other than the miner's wages.
opencast a quarry in which overlying rock layers are removed to reach the coals.
outcrop where a rock type is present beneath the soil or glacial sand and gravel.
oversman an overseer in a mine, above the Deputy but below the Under Manager.

P

parrot coal a poor cannel coal used for gas-making. Chatters when burned.
parting a plane along which a uniform rock splits into several layers.
pavement the rock floor beneath a coal seam.
pelt a bituminous shale often bearing plant debris.
piece poke miner's lunch box.
pillar column of undisturbed rock between worked rooms in a coal seam.
pit underground workings connected by a shaft.
plies rapid alternations of interbedded hard and soft rocks.
ply a layer of hard rock separated from its neighbours by partings.

R

rank the degree of alteration from peat to anthracite due to loss of volatiles.
redd to clear away waste or debris in workings or roads.
rib thin layer of hard material e.g. stone layer in a coal seam.
road a passageway in an underground working area.
roof the rock layer above a coal seam.
room an opening in a coal seam within which the hewer cuts the coal.
rough coal a free coal associated with splint or cannel but lower in quality.
rum carbonaceous, silty mudstones, formerly used in lime burning.

S

sclit shaly coal or coal-bearing blaes.
seatearth the foundation layer beneath the coal.
shale laminated mudstone.
sill body of intrusive igneous rock, gently inclined or parallel to bedding.
splint hard, coarse coal that breaks unevenly and travels well. Gives great heat.
stoop and room method of mining by cutting rooms separated by pillars or walls.

T

tirr, tirring to remove detritus from the rock layer above the coal seam.
trouble disturbance of coals by faulting, folding or baking by igneous intrusion.

W

waste rubble from old workings, often still below ground.
winning the entire process of releasing coal from a seam by quarrying or mining.

Bibliography

Adams, P.J. (1980) *The Origin and Evolution of Coal*. London: HMSO.

Anderson, J. (1859) *Dura Den: a Monograph of the Yellow Sandstone and its Remarkable Fossil Remains*. Edinburgh: Constable.

Arnot, Robert Page (1955) *A History of the Scottish Miners*.London: Allen & Unwin.

Ashton, T.S. and Sykes, J. (1929) *The Coal Industry in the Eighteenth Century*. Manchester: Manchester University Press.

Bald, R. (1812) *A General View of the Coal Trade of Scotland*. Edinburgh: Oliphant, Waugh and Innes: Google ebook.

Banfield Coal Book (1761) in Cupar Library Local History Collection.

Barrowman, J. (1886) *A Glossary of Scotch Mining Terms*. In the Scottish Mining Website, http://www.scottishmining.co.uk/Indexes/Barrowman.html (accessed 6 January 2016).

Barrowman, J. (1897–8) Slavery in coal mines of Scotland. *Transactions of the Federated Institution of Mining Engineers*, Vol. XIV. In British Geological Survey Library, Riccarton, Edinburgh.

Beaumont, F. (1806) *Report on the Ceres Coals*. In Forbes, S.B. (1955) Unpublished notes in British Geological Survey Library, Riccarton, Edinburgh.

Bell, Rev. A. (1790) *Crail: First Statistical Account*, Vol. 9, 1791–99, No. XXX, pp. 439–458.

Belt, E.F. (1975) Scottish Carboniferous cyclothem patterns and their paleoenvironmental significance. In Broussard, M.L. *et al.* (1981) *Delta models for exploration*. Texas: Houston Geological Society pp. 427–429.

Birley, R. (2009) *Vindolanda: a Roman Frontier Fort on Hadrian's Wall*. Stroud: Amberley Publishing Co..

Bluck, B.J. (2002) The Midland Valley Terrane. In Trewin, N.H. (2002) *The Geology of Scotland* (4th edn). London: Geological Society of London, pp.149–166.

Brodie, Rev. A. (1791) *Carnbee: First Statistical Account*, Vol. 13, 1791–99, No.III, pp. 23–33.

Bruce, W. Scott. (1980) *The Railways of Fife*. Edinburgh: Mercat Press.

Burn, M.J. (1990) *Controls on the Geometry of Carboniferous Deltas: East Fife, UK and East Kentucky, USA*. Unpublished thesis, Oxford Polytechnic.

Children's Employment (Mines) Commission. (1842) Parliamentary Papers, Vols XVI–XVII.

Cleal, C.J. and Thomas, B.J. (2009) *An Introduction to Land Fossils*. Cambridge: Cambridge University Press.

Cobbett, William (ed. Ian Dyck, 2001) *Rural Rides*. London: Penguin Books.

Creighton, C. (1965) *History of Epidemics in Britain* (2nd revised edn). London: Frank Cass and Co. Ltd.

Cunningham, A.S. (1913) *Mining in the Kingdom of Fife* (2nd edn). Edinburgh: Romanes & Son.

Davies, J. (1993) *A History of Wales*. London: Penguin Books.

Dott (1944) *The East Fife Coalfield* [MSS]. Land Survey Archive, LSA 238. British Geological Survey Library, Edinburgh.

Davydov, V.I., Korn, D. and Schmitz, M.D. (2012) The Carboniferous Period. In Gradstein, F.M., Ogg, J.G., Schmitz, M.D. and Ogg, G. (eds) (2012) *The Geologic Time Scale* Vol. 2. Amsterdam: Elsevier pp. 603–652.

Dowden, J. (ed.) (1903) *The Chartulary of the Abbey of Lindores*. Edinburgh: Scottish History Society.
Dron, R.W. (1902) *The Coalfields of Scotland* (1st edn 368pp; 2nd edn 1921). Glasgow: John Smith.
Duckham, B.F. (1968) Life of Laborers in a Scottish Colliery, 1698–1755. *Scottish Historical Review* (47) pp. 109–128.
Duckham, B.F. (1970) *A History of the Scottish Coal Industry*, Vol. 1, 1700–1815. Newton Abbot: David & Charles.
Fielding, C.R., Al-Rubil'i, M. and Walton, E.K. (1988) Deltaic sedimentation in an unstable tectonic environment: the Lower Limestone Group, (Lower Carboniferous) of East Fife, Scotland. *Geological Magazine* (125) pp. 241–255.
Forbes, S.B. (1955) *The Coalfields of Fife*. Unpublished Report [MSS] for the National Coal Board, 4 vols with maps and sections.
Fordyce, W. (1860) *Coal, Coke and Coalfields and Iron Manufacture in Northern England*. London: Sampson, Low & Son.
Forsyth, I.H. and Chisholm, J.I. (1968). Geological Survey boreholes in the Carboniferous of East Fife, 1963–64. *Bulletin of the Geological Survey of Great Britain* (28) pp. 61–101.
Forsyth, I.H. and Chisholm, J.I. (1977) *The Geology of East Fife*. Memoir of the Geological Survey of Great Britain. Edinburgh: HMSO.
Francis, E.H. and Hopgood, A.M. (1970) Volcanism and the Ardross Fault, Fife, Scotland. *Scottish Journal of Geology* (6) pp.162–185.
Francis, E.H. (1991) Carboniferous. In Craig, G.Y. (ed.) *The London Geology of Scotland* (3rd edn) London: Geological Society Publishing, pp. 347–392.
Galloway, R.L. (1882) *A History of Coal Mining in Great Britain*. London: Macmillan & Co.
Geikie, A. (1900) *The Geology of Central and Western Fife and Kinross*. Memoir of the Geological Survey of Scotland. Glasgow: HMSO.
Geikie, A. (1902) *The Geology of Eastern Fife*. Memoir of the Geological Survey of Scotland. Glasgow: HMSO.
Gibson, W. (1927) *Coal in Great Britain*. London: Edward Arnold & Co.
Goodlet, G.A. (1957) Lithological variation in the Lower Limestone Group in the Midland Valley of Scotland. *Bulletin of the Geological Survey of Great Britain* (12) pp. 52–65.
Graham, F. (1979) *The Roman Wall*. Newcastle-upon-Tyne: Frank Graham.
Grange Colliery Output Book: Daily Accounts and Weekly Summaries 1826–1830. Cupar Library, Cupar, Fife. [Shelf Mark: L622.33].
Grant, I.R. and Withrington, D.J. (eds) (1978) *Third Statistical Account of Scotland, Vol. X: Fife*. EP Publishing Ltd.
Greensmith, J.T. (1965) Calciferous Sandstone Series sedimentation in the eastern end of the Midland Valley of Scotland. *Journal of Sedimentary Petrology* (35) pp. 223–242.
Hooper, M.D., Davies, S.J., Underhill, J.R., Monaghan, A. and Browne, M.A.E. (2002) *Evolution of the Carboniferous Central Coalfield of the Midland Valley of Scotland and Implications for the Regional Tectonic Setting. Abstract of Conference on Hydrocarbon Reserves of the Carboniferous Southern North Sea and Surrounding Onshore Areas.* (59) Leeds: Yorkshire Geological Society.
Kassi, A.M., Weir, A.J. and Browne, M.A.E. (1998) The stratigraphy and depositional context of a temporary exposure in the Anstruther Formation (Strathclyde Group), Anstruther Wester, Fife. *Scottish Journal of Geology* (34) pp. 127–132.
Kassi, A.M., Weir, J.A., McManus, J. and Browne, M.A.E. (2004) Lithofacies and sedimentary cycles within the Late Dinantian (Late Brigantian) of Fife and East Lothian: Is a sequence stratigraphical approach valid? *Transactions of the Royal Society of Edinburgh: Earth Sciences* (94) pp. 95–113.
Keddie, Henrietta [Sarah Tytler] (1911) *Three Generations: the Story of a Middle-class Scottish Family*. London: John Murray.

Kerr, R.D. (ed.) (1980) *A Glossary of Mining Terms used in Fife*. Published by Fife Colleges.
King, I. (2000) *Thornton Railway Days*. Kelty, Fife: Windfall Books.
Kirk, S.R. (1925) *A Coastal Section in the Calciferous Sandstone Series of Eastern Fife*. Unpublished PhD thesis, University of St Andrews.
Kirkby, J.W. (1880) On the marine fossils in the Calciferous Sandstone Series of Fife. *Quarterly Journal of the Geological Society* (36) pp. 559–590.
Lamont, J. (1810) *The Chronicle of Fife: Diary of John Lamont of Newton, 1649–1672*. National Library of Scotland, Edinburgh,
Landale, D. (1835) Report on parts of the coal district situated between the Forth and the Tay. *Transactions of the Royal Highland Agricultural Society* X, pp. 410–439.
Landale, D. (1837) Report on the geology of the East Fife Coalfield. *Transaction of the Royal Highland Agricultural Society* XI, pp. 265–348.
MacGregor, A.R. (1968) *Fife and Angus Geology: an Excursion Guide*. Edinburgh: Blackwood & Sons.
McKechnie, J. Macgregor. (1958) *A Short History of the Scottish Coal Mining Industry*. National Coal Board, Scottish Division.
McManus, J. (2010) *Mining between Ceres and St Andrews*. Kelty, Fife: Windfall Books.
Major, John (1521) In Hume Brown, P. (1893) *Scotland before 1700 from Contemporary Documents*. Edinburgh: D. Douglas, pp. 58–59.
Martin, Colin (1999) *The 18th Century Industrial Landscapes between St Monans and Pittenweem: a Cartographic and Archaeological Study*. Dundee: Abertay Historical Society, Monograph 2, pp. 42–65.
Martin, M., Sparling, C. and Martin, C. (2007) *Fife the Kingdom of Coal*. A Fife Pits Production. Fife Pits Website: www.fifepits.co.uk.
Martin, Paula (1999) *The Newark Coal and Salt Work Company: its Social and Economic Impact*. Dundee: Abertay Historical Society, Monograph 2, pp. 28–41.
Merson, Rev. William (1845) *Second Statistical Account for the Parish of Crail*. Vol. 9, pp. 941–969.
Millar, John (1773) *Observations Concerning the Distinction of Ranks in Society*. London: John Murray.
Minerals Planning Guidance 3 (1999) *Coal Mining and Colliery Spoil Disposal*. London: HMSO.
Montgomery, G. (1994) *A Mining Chronicle*. Musselburgh: Newcraighall Heritage Society.
Muir, A. (1958) *The Fife Coal Company: a Short History*. Leven: The Fife Coal Company Ltd.
Muirhead, Rev. G. (1792). *Dysart: First Statistical Account of Scotland*. Vol. 9, No. XXXV, pp. 502–524.
Parry, H.J. (1985) *Coping with Fife's Mining Industrial Heritage*. Fife Regional Council.
Read, W.A., Browne, M.A.E., Stephenson, D. and Upton, B.G.J. (2002) Carboniferous. In_ Trewin, N.H. (ed.) *The Geology of Scotland* (4th edn). London: Geological Society of London. pp. 251–299.
Scottish Mining Website: http://www.scottishmining.co.uk/Indexes/Barrowman.htmlSibbald, Sir Robert (1803) *The History, Ancient and Modern, of the Sheriffdoms of Fife and Kinross, with Descriptions of Both, and of the Firths of Forth and Tay and the Islands in them* (2nd edn). Edinburgh: James Watson.
Silver, O. (1987) *The Roads of Fife*. Edinburgh: John Donald.
Smith, Adam. (1776) An Inquiry into the Nature and Causes of the Wealth of Nations. London: Strahan and Cadell.
Smout, T.C. (1964) *A History of the Scottish People 1560–1830*(1998 edition). London: Fontana Press.
Sparling, C. (2004) *The Oncost Book of a Fifeshire Colliery: Grange Colliery, Earslferry, 1826*. Fife: Chris Sparling.
Sparling, C. (2005) *Coal Working around Crail*. Fife: Fife Pits Production.

Sparling, C. (2008) *Buried Sunshine*. Unpublished compilation.

Stansfield, G. (1998) *Fife's Lost Railways*. Ochiltree, Ayrshire: Stenlake Publishing.

Tait, D. and Wright, J. (1923) Notes on the Structure, Character and Relationship of the Lower Carboniferous Limestone of St Monans, Fife. *Transactions of the Edinburgh Geological Society* (11) pp. 165–184.

Taylor, J. (fl. 1705)(1903) *A Journey to Edenborough in Scotland*. Revised and edited by William Cowan 1903. Edinburgh: W. Brown.

Terris, I. (undated) Coal Mining in the East Neuk of Fife. Unpublished report, copyright Ian Terris, 32pp.

Terris, I. (2010) *Coal Mining in the East Neuk of Fife*. Unpublished report: copyright, Ian Terris.

Towill, Edwin Sprott (1983) *Saints of Scotland* (2nd edn). Edinburgh: St Andrew Press.

Traquair, R.H. (1901) Notes on the Lower Carboniferous fishes of Eastern Fife. *Geological Magazine* (4) pp. 110–114.

Trewin, N.H. (2002) *The Geology of Scotland* (4th edn). London: Geological Society of London.

Trewin, N.H. and Thirlwall, M.F. (2002) Old Red Sandstone. In Trewin, N.H. (ed.) *The Geology of Scotland* (4th edn). London: Geological Society of London, pp. 149–166.

Underhill, J.R., Monaghan, A. and Browne, M.A.E. (2000) *Controls on Carboniferous Basin Development and Petroleum Prospectivity in the Midland Valley of Scotland. Abstract of Conference on Hydrocarbon Reserves of the Carboniferous Southern North Sea and Surrounding Onshore Areas.* (66) Leeds: Yorkshire Geological Society.

Vail, P.R. and Mitchum, R.M. Jr. (1977) *Seismic Stratigraphy and Global Changes of Sea Level.* American Association of Petroleum Geologists, Memoir 26 pp. 145–163.

Van Wagoner, J.C., Posamentier, H.W., Mitchum, R.M., Vail, P.R., Sarg, J.F., Loutit, T.S. and Handenbol, J. (1988) An Overview of the Fundamentals of Sequence Stratigraphy and Key Definitions. In Wilgus, C.K. *et al.* (eds). Sea Level Changes: an Integrated Approach. *Society of Economic Palaeontologists and Mineralogists*, Special Publication 42 pp. 39–45.

Wattison, A. (1962) Temporary exposures in the Calciferous Sandstone Measures of East Fife. *Transactions of the Edinburgh Geological Society* (2) pp. 119–124.

Weibel, C.P. (1996) Application of sequence stratigraphy to Pennsylvanian strata in the Illinois Basin. In Witzke, B.J. *et al.* (1996) *Paleozoic sequence stratigraphy: views from the North American Craton*. Geological Society of America, Special Paper 306, pp. 332–339.

Whatley, Christopher A. (1984) *'That important and necessary article': the salt industry and its trade in Fife and Tayside, c.1570–1850*. Dundee: Abertay Historical Society.

Whatley, Christopher A. (1987) *Scottish Salt Industry 1570–1850: an economic and social history*. Aberdeen: Aberdeen University Press.

Unpublished Reports and papers held in the Library of the British Geological Survey, Riccarton, Edinburgh

Barrowman (1886)
Dixon and Ball (1808)
Geddes (1854)
Gemmell (1905)
Grieve (1830)
H. Keddie (1922)
Maxton (1848)
Sanderson (1982)
Williamson (1839)

Index

Page numbers in *italic* denote figures. Page numbers in **bold** denote tables.